CROSS CULTURAL WORKSHOP WITH WOODEN CONSTRUCTION

国际木构工作营

赵辰　主编

中国建筑工业出版社

图书在版编目（CIP）数据

国际木构工作营/赵辰主编. —北京：中国建筑工业出版社，2008
ISBN 978-7-112-09852-1

Ⅰ.国…　Ⅱ.赵…　Ⅲ.木结构-高等学校-教学参考资料　Ⅳ.TU366.2

中国版本图书馆CIP数据核字（2008）第041559号

南京大学建筑学院研究生教学中的建构研究，近年来在木构建造方面有显著的发展。本书记录了由南大建筑学院和挪威奥斯陆建筑与设计学院共同合作的“国际木构工作营”的工作过程：在南京由挪威学生与中国学生及相关教师完成了两个木构物；随后在挪威的特维德斯特兰德的“挪威森林”中建造了五个木构物。从理论研究加模型制作探讨发展到了在真实场景之中的建造，全过程都在教师以及木工师傅指导下由两校学生共同参与完成。本书可供高等学校建筑学、城市规划、风景园林、艺术设计等专业的师生教学、研究之用，也可作为广大专业人士学术交流和工程实践的参考用书。

责任编辑：陈　桦
责任校对：陈晶晶　安　东

CROSS CULTURAL WORKSHOP WITH
WOODEN CONSTRUCTION
国际木构工作营
赵辰　主编
*
中国建筑工业出版社出版、发行（北京西郊百万庄）
各地新华书店、建筑书店经销
北京嘉泰利德公司制版
北京方嘉彩色印刷有限责任公司印刷
*
开本：889×1194 毫米　1/20　印张：$5\frac{4}{5}$　字数：260 千字
2008 年 6 月第一版　2008 年 6 月第一次印刷
印数：1-2500 册　定价：42.00 元
ISBN 978-7-112-09852-1
(16556)

2006 09-2006 11

南京大学建筑学院
School of Architecture,NJU
奥斯陆建筑与设计学院
School of Architecture and Design in Oslo(AHO)

主编：赵辰
Editor in chief：Zhao Chen

编辑：冯金龙　周凌　程云杉　沙米・林塔拉
Editor：Feng Jinlong Zhou Ling Cheng Yunshan Sami Rintala

版面：陈永乐　侯博文　罗辉
Graphic：Chen Yongle Hou Bowen Luo Hui

配文：罗丹青　彭嫱　侯博文　蔡伟森　陈永乐　尤伟
Text：Luo Danqing Peng Qiang Hou Bowen Cai Weisen Chen Yongle You Wei

目录
Contents

国际木构工作营 · 前言

2007. 1，南京
赵辰
南京大学建筑学院教授

南京大学建筑学院研究生教学中的建构研究，近年来在木构建造方面有显著的发展。从理论研究加模型制作探讨发展到了在真实场景之中的建造，全过程都有学生和教师参与。2006 年，在原有的木构建造教学的基础上，整个教学程序又发展到新的阶段。由南大建筑学院和挪威奥斯陆建筑与设计学院共同合作的"国际木构工作营"被成功地运作了。"工作营"由两个阶段组成：第一阶段，于 2006 年 9 月在南京，十七名挪威学生与两位教师来到南京大学，与十二名中国学生及相关教师汇合，在一个星期之内，完成了两个木构物的现场重新安装。它们是坐落在南大浦口校园的"桥"和"亭"，作为景观建筑基本概念的"移动"(Moving)、"停留"(Staying)之体现；在 11 月，我们再次在挪威的小镇特维德斯特兰德聚合，开始第二阶段的工作营。在当地业主的协助下，在所谓"挪威森林"中建造了五个木构物："攀爬架"、"迷宫"、"观星台"、"树屋"和"地表延展"，它们是为小镇的小孩运动公园提供各种场所。

工作营带给我们许多重要的建筑教学与研究方面的经验，基本可以分为以下几个概念："建造"、"木材"以及"国际工作营"。

关于建造

由于今天建筑设计的社会定义已经被多数人接受，我们几乎忘却了在建筑(Architecture)成为一种社会分工和学术科目之前，建造在建筑中是一切的一切。设计原本也包含在建造之中。在当代的建筑生产程序中，建造依然是复杂的设计过程之终结者。作为建筑的作品，必然将由建造来最终传达。因此，在建筑研究和教学中考虑建造总是极为重要的，尽管大多数情况下建筑学院中的设计停留在纸和屏幕上。

1920 年代的德国包豪斯，开始在现代建筑教育中融入建造，其理念是将工艺传统在实用艺术中实现。其中典型的事例是，由格罗皮乌斯（Walter Groupius, 1883-1969）主持的学生工作营在柏林成功地建造了萨默菲尔德别墅(1920-1921)，那是一次传统式样的木构建造。作为一极其重要的建构意义上的建筑学理解，建造活动已经在今天世界建筑学院中成为相当普遍的方式。但是，有时会被研究模型，尤其是一比一的足尺模型所混淆。我们应该将教学中的建造明确定义为两个基本条件：一是"真实的尺度"；二是"真实的材料"。这意味着教学中的建造，需要同时满足真实人体尺度和真实建筑材料这两个条件。以此，建筑学的研究将真诚地关注：对材料的感知，光感、色彩以及节点的象征与触觉意义。

关于木材

从理论上讲，建筑用材料是无限制的。人类自古以来运用任意材料来建造；只是今天的建筑工业体系中的常用建筑材料是有一定限制的。在现实的建筑教学之中，我们知道被运用的是少数的几种建筑材料：例如砖、石材、混凝土，还有木材。很显然，木材在学生的工作中是更合理和理想的建造材料。由于在模型材料中，木材已经显示其优点：不仅在于便于操作、加工，以及购买与运输方便；同时木材也在建造活动中充分地显示出材料的真实感。在我们的案例中，木材帮助我们理解真实的木构并且是所有细部的连接。木材再次向我们展示了其建造的优点；以其加工、运输的方便，更以对体系化的设计和建造之影响，并有可能在全过程中装配与重新装配。

关于国际工作营

国际间的工作营常常在世界上的建筑学院受到欢迎：首先，这意味着双方学生的学习热情容易被激发，正是由于对对方文化的兴趣；其次，将大大鼓励设计与建造过程中的探索性，由于给予不同文化背景的观点和观念产生了交流和共享。在这次的木构工作营中，不论是在南京还是在特维德斯特兰德，中国学生和挪威学生各自以其不同的文化背景，互相交流了他们的设计概念、建造方式，以及生活经验。不可避免地有大量的聚会、讨论、争辩，以及幽默和笑声。他们非常成功地在工作中发展了合作对象，在生活中成了朋友。在南京工作营，挪威学生得到了中国学生的大力协助而充分了解了南京城以及中国文化。而在宁静的挪威小镇特维德斯特兰德，中国学生在对方的帮助下，学会了在欧洲的生活和相关的文化。可以肯定地是，这种人生经历将深刻地影响他们未来的工作和生活。

几乎在工作营的第一天，中国学生已经发现他们的挪威合作者们在木建造方面要明显更具有能力和经验，在后来的工作过程中挪威学生的作用得到了积极的发挥，而中国学生的努力和勤奋使他们的木构工艺与对建造的理解都有极大的提高。很显然，这种跨文化的工作营，学生们能够从中得到的建筑与设计之知识与体验，是完全不可能在普通的工作室和课堂中得到的。

尽管在建造活动中学生们遇到的困难是不断的，为此，他们必须为之花费巨大，几乎可以以“损失”来形容。 但是，以建造工作营是典型的“实践中学习”之教学方法，也同时必然印证另一句名言：“有得必有失”。

Nan Jing, January, 2007
Zhao Chen
Professor SA

Under the existing framework of tectonic studies in the post-graduate program, the School of Architecture in Nanjing University has developed construction work, especially with wood, in recent years. It moved from theoretical research with exploring model making to the realistic construction in site, whole process is involved by students and teachers. In 2006, based upon the experience of wood construction in studio of last year, the program has been successfully developed to a new stage, an international student workshop lunched with the cooperation between the School of Architecture, Nanjing University and the School of Architecture and Design in Oslo (AHO) in Norway. The workshop formed by actually two phases, the first one acted in September 2006, Nanjing, 17 Norwegian students headed by two teachers from AHO, together with 12 students and some teachers from NJU, within one week, we resembled two objects of wooden construction, a “Bridge” and a “Pavilion” corresponded to concepts of “Moving” and “Staying”, at the Pukou Campus of NJU. Then, in November, we gathered again at Tvedestrand, Norway, to run the workshop in second phase, with local client to construct five objects of “Climbing”, “In and Out”, “Stargazing”, “Tree House” and “Play Ground” in the Tvedestrand Aktivetpark of “Norwegian woods”.

The workshop brings to us crucial experience in architectural teaching and study, which can be represented as three concepts of “Construction”; “Wood” and “Cross Cultural Workshop” as following:

On Construction

Since today’s social definition of architecture design is accepted by the most people, we almost forget the fact of that construction had been everything in building process, before the architecture became a kind of academic discipline in school and a profession in society. Design should be included in the process of construction. Even in contemporary produce program of architecture, construction would be the terminator of complicated design process. As a works of architecture, should be at last delivered by construction. Therefore, it should be always important to consider about the construction in architectural studies and teaching, however in most cases architectural school spend more time in design with paper or screen.

Bauhaus in 1920’s, started to merge construction into the modern architecture education, just in meaning of to realize the craftsman tradition in applied arts. Typically, Walter Groupius (1883-1969) conducted a student workshop to build Sommerfeld House (1920-1921) in Berlin, which was a wooden construction in traditional style. Construction, as the important process for students to understand architecture concerned with tectonic meaning, becomes pretty popular way to be developed in today’s architectural schools in the World. However, it would be sometime confused with modeling, especially when the models with real size of one to one scale. We shall define construction in studio with both elements of “real size” and “real material”, which

means construction in studio should be realistically related to human scale and building materials. With it, architectural study would be authentically concerned in the sensuous qualities of materials, light, and color, and in the symbolic, tactile significance of the joint.

On Wood

Theoretically, building materials are unrestricted. Human happened to utilize any different material in construction historically; however there is certain limitation of popular materials in today's building industry. To be realistically in architectural study, there are basically existed few kinds of building materials one can see: they are bricks, stone, concrete as well as wood. Nevertheless, we find wood should be more than reasonable and ideal material to be applied in studio for students work. Since wood as the model material already shows the advantages of easy taking and working, cheap to have and for transport, but also shows real material feeling. In our case, the wood material helped us to understand exact wooden construction in very detailed joints. Wood shows again even advantage and quality in the construction practice; because of the convenience in manufacturing, transporting, further more with the systematic design and construction, it is possible to assemble and dissemble within the whole process.

On Cross Cultural Workshop

Cross cultural workshop has been often welcome by many architectural schools in the World: The first of all, it would motivate the passion of students from both sides, since curious of other cultures; Secondly, it would encourage the exploring in the process of design and construction, since different views be exchanging and sharing. With this wooden construction workshop, in Nanjing or in Tvedestrand, Chinese and Norwegian students exchanged each other with design concepts, construction ways, also the life experience, certainly based upon their culture backgrounds. There were inevitable meetings, discussions, arguments, and humors, laughing; they developed very well their team-mates in working, and friends in their life. When the workshop in Nanjing, students from Norway were helped by Chinese students to know the city and Chinese culture very detailed. While in Tvedestrand, a quite and small town at the south coast of Norway, with helps of locals, students from Nanjing learned to life and understand the European culture. It is sure that, those experiences will influence very much them for their life and profession in the future.

Chinese students found their Norwegian cooperators are much capable and experienced in wooden construction almost at the first day of the workshop, whatever boy or girl. So that on the following days, Norwegian students continuously helped the construction work positively; with the diligence and hardworking, Chinese students developed their capability in construction and specifically on woodcrafts.

It is clear that, with this kind of cross cultural workshop, students learned and experienced so much in architecture and design, which certainly is impossible to be gained in normal studio or class.

However, there is no easy way to pass in construction, so students have to spend a lot in the workshop, even should be "pain" in certain meaning. But, as the construction workshop is sort of speaking as "learning by doing", it should be also demonstrated as another saying of "no pain, no gain".

亭 Pavilion
桥 Bridge
基地 Site
南京 Nanjing
塘
花房

南京
NANJING
2006.03-2006.09

从南京到特维德斯特兰德森林：国际木构工作营

2007.7，南京
周凌
南京大学建筑学院副教授

南京大学建筑学院与奥斯陆建筑学院国际联合木构工作营，于2006年9月开始，至2006年11月底结束。活动分为两个阶段：第一阶段工作营地点在南京，奥斯陆建筑学院一行29名同学与南京大学12名研究生在南京大学浦口校区搭建一个木结构亭子和一座木结构桥；第二阶段南大学生前往挪威，在挪威南部风景优美的小镇特维德斯特兰德（Tvedestrand），与挪威学生一同在紧靠峡湾的公园内设计建造五个木结构建筑："树屋"（Tree house），"攀爬架"（Climbing），"迷宫"（In and out），"地表延展"（Playground），"观星台"（Stargazing）。

第一阶段（南京阶段）的活动是南京大学建筑学院"材料与建造"课的延续，这个课程每年春季开课，对象是一年级硕士研究生，目的是训练学生动手能力与建构思维。针对国内建筑教育重视图面而轻视建造的现象，课程关注重点是材料、结构、构造、节点、建造等技术性问题。这些问题不仅影响建筑的实际使用，也影响形式，是最终形式的重要组成部分。2006年建构课由冯金龙教授与周凌副教授共同担任，课程前期进行四个真实比例的公园小品设计建造。其中一个候车廊、一个休息平台、一个亭子和一座木拱桥。另外一门重要的课程是赵辰教授开设的"中国木建构文化研究"，课程中不仅有关于中国传统木构文化的理论研究，也包括一些木工基础训练，如做木凳，制作"六木同根"等，通过亲身体验，对木构开始有所理解。在两门课三个老师共同指导下，完成了最后的1 ：1真实木构建造。在本校建构课结束后，南大建筑学院选派12个同学，与奥斯陆建筑学院的同学组成联合工作营，于2006年9月，用一周时间在南京大学浦口校区搭建两个木构建筑——木亭与木桥。

第二阶段（挪威阶段）的活动是在挪威的公园中建造木构小品，中方学生一行12人于2006年10月飞往挪威，开始为期两周的建造活动。两国学生为小镇的开放式公园工作，公园三面环山，一面是海，五个建筑小品就是为当地的孩子和残疾人设计的。峡湾（fjord）是挪威举世闻名的地貌，海水深入陆地，两侧是犬牙交错的陡峭山崖，特维德斯特兰德小镇正是因其峡湾而得名。中方带队老师周凌副教授、程云杉博士与工作营同学一道，每天从小镇步行到峡湾公园山顶工地，配合同学并解决一些技术问题。这次活动的赞助商有两个：一个是小镇的镇委会，一个是小镇上的一家大型轮船设备制造企业。碰巧的是，这家企业在中国的分厂正好设在南京，产品一部分提供亚洲市场。镇委会的几个工作人员每天到现场积极配合施工。最终完成的五个小品各具特色。

进入公园，第一个看到的便是由陈永乐、柳巍等人和挪威学生建造的"树屋"——在橡树林中选择3棵大树当支撑柱，平台和房子就搭在离坡地7m高左右的空中。由于是悬空操作，未知因素很多，两国学生在设计上只花了一天时间。中国学生与挪威学生设计理念差异很大，中国学生认为需要再加些支撑柱，保证"树屋"的稳定，但挪威学生却反对任何添加，他们觉得"树屋"随着树一起晃动，甚至随着树的生长慢慢长高才是最贴近自然的。"树屋"里还专门加了床，学生们认为睡在里面听鸟叫的感觉一定很美。在具体的建造中，安装平台所需要的3根木梁居然花费了整整3天时间。在去挪威之前学生们从未想过要在树上工作，由于没有大型施工机械，自己做了梯子，测定距离的办法是爬到梯子顶端拉起绳子量，为了把沉重的木梁拉上去，他们还专门制作了简易滑轮。"树屋"的对面是"攀爬架"。"攀爬架"的构思是，几个3m见方的立方体空中叠加，每层平台向不同方向出挑，像一个有

动感的向上生长的构筑物。这个小品实施相比之下最顺利，立柱搭好，再搭每层平台楼板，之后重复叠加就行。技术难度不大，主要依靠数量累积取得效果。

“攀爬架”旁边的缓坡地便是“迷宫”的基地。“迷宫”的构思是，森林中相对平缓的坡地上，建造一组可以旋转的木板墙体，形成一个内外空间可以相互转换的活动场地，孩子们通过旋转木板来自己组合空间，他们会发现一个有趣的现象，经过旋转，一些外部空间突然会变成内部空间。建造最初构思旋转依靠轴承实现，后来因为加工、维护等原因，改为比较简单的双层套管旋转构造。“地表延展”位于“迷宫”旁边的陡峭坡地。“地表延展”的构思是，在森林陡峭的坡地上顺应起坡方向，制作一组起伏的地板，像自然地形起伏一样，孩子们可以在上面走动、游玩、休息。在搭建细节中，两国学生充分考虑到了儿童和残疾人的需要，比如地面木条间的缝隙宽度，不能太窄——落叶卡在缝隙中结霜会使孩子滑倒，不能太宽——不至于让轮椅的轮胎卡在缝隙中，要让它们顺利通过。

“观象台”位于整个基地的最高山头，是建造难度最大，也是最特殊的一个基地。“观象台”风景最美，视野开阔，引人入胜。当初学生们的考虑是在山顶建造标志性建筑，但在一起商讨后认为，“观星台”建造的目的是为了防止昂贵的天文望远镜被淋湿、被盗，有一个覆盖物就可以了，设计简化到最后，就是一个一人高的木盒子，加上一块4m×8m的木平台。设计时，考虑到观测者休息的需要，特意在木盒中安排了一个人的坐椅。实际上，木盒子本身既是望远镜的遮蔽物，又是一只带顶的大椅子。更巧妙的是，为了满足更多观测者的参与，这只木盒可以在平台上移动，人多的时候把盒子挪开，人们就能很方便地在望远镜旁围观。

经过两周工作，五个建筑小品完成了主要结构部分，其中观象台、旋转迷宫、攀爬架、游戏场全部完工，已经可以投入使用。树屋因为建造难度过大以及其他一些技术原因没有完成上部结构，镇委会的工作人员表示他们将继续完成。建造活动结束时，当地的小孩子们已经成为第一批使用者，他们在攀爬架、地表延展上快乐地玩耍，好奇地旋转着木屏风。挪威虽然是一个十分发达、现代化的的国家，但人们的传统家庭观念很强，多数家庭都有一两个孩子。由于人们都喜欢小孩，便自然都关注这些大型“玩具”，家长们常常背着孩子来工地，参观建造活动，了解进度。当地一些报纸也进行了报道，甚至有一些其他小镇的家长也因为看到报道专程带孩子来参观建造活动。联合工作营顺利完成了主要木建筑的建造，完成了建造活动，两方同学的工作，得到了当地镇委会的大力赞赏。尤其是中方同学，刚开始对木工工具与技术都不熟练，但他们勤奋学习，最终得到了镇委会工作人员的再三肯定和褒奖，当地人十分感谢这些来自远方国度的勤奋而聪明的青年们。

更有意义的是，通过亲身建造体验，学生们充分体会到了材料特性、材料构造方式及其表现形式以及尺度等问题，体验到了材料的重量，理解了重力对于建筑学的意义，不是纸上画画就能成为建筑师。木材作为主要的建筑用材之一，其特殊的材料性能、构造和结构方式体现了丰富的意义和内涵。材料是建筑的物质基础，材料的合理选择和使用是一个基本问题，此次建构教学基于对材料基本的物理力学性能、使用特性、连接方式以及建造技术特点等，通过实际建造的过程，帮助学生们思考由材料、结构和构造方式所形成的木构建造的逻辑关系，理解形式产生的物质与技术基础。

From Nanjing To Tvedestrand:International Workshop For Wooden Construction

Nan Jing, July, 2007
Zhou Ling
Associate Professor SA

The International Workshop for Wooden Construction of School of Architecture of Nanjing University(SA NJU) and Oslo School of Architecture began in Sept., 2006 and ended at the end of Nov., 2006, of which the activities included two parts: in part 1, all the activities were conducted in Nanjing, where 29 students of the Oslo School of Architecture and 12 overgraduates of Nanjing University built a wooden gloriette and a bridge in the Pukou campus of Nanjing University. In part 2, the students of Nanjing University came to Norway, together with the Norwegian students, designed and built five wooden buildings in the park close to the fjord of Tvedestrand, a small town in Southern Norway with very beautiful landscaping, namely Tree house, climbing, In and Out, Playground and Stargazing.

The activities in the first stage (Nanjing Stage) were the continuation of the course on Material and Construction held by the School of Architecture of Nanjing University. Such course, targeting grade one overgraduates, begins in the spring of each year for the purpose of training the manipulation capacity and new concept building of students. As the domestic architectural education thinks much of the architectural drawing and makes light of the construction, such course focuses on the technical issues relating to materials, structure, construction, joint and building. These issues relating to the influence upon both the practical utilization and the form of the building serve as the key part of the final form. In 2006, Professor Feng Jinlong and Associate Professor Zhou Ling were assigned to teach the architectural construction course jointly. During the whole course, the prophase was related to the design and construction of the four works in a park (in real size), including one waiting room, one resting platform, one gloriette and one wooden arch bridge. Another important course is the Study on the Culture of the Wooden Architectural Construction in China and Professor Zhao Chen is assigned to teach this course. Such course not only covers the theoretical research on the culture of the traditional wooden construction in China but also includes some foundamental architectural training (i.e. making wooden stool and building Liu mu tong geng, etc.). Through practical experience, the students may have a basic understanding to the construction course. As for these two courses, all the participants completed the final true wooden construction with a scale of 1:1, under the instruction of three teachers. At the completion of the architectural construction course, 12 students of SA NJU were selected and constituted a joint workshop in cooperation with the students of Oslo School of Architecture. In Sept., 2006, the joint workshop built two wooden architectural buildings in the Pukou campus: a wooden gloriette and a wooden bridge.

The activities in the second stage (Norway Stage) were to build wooden construction works in a park in Norway. In Oct., 2006, a dozen Chinese students flew to Norway and commenced the construction that lasted two weeks. The students from both nations worked for the open park of the town named Tvedestrand. This park is surrounded on three sides by mountains and one side by sea, in which five buildings are designed for local children and the disabled. The fjord is a worldwide well-known scenic spot in Norway, where the seawater penerates into the continent and the cragged cliffs are interlocking on both sides. The fjord alone makes the small town of Tvedestrand famous. Professor Zhao Chen and Assistant Professor Zhou Ling, the leading teachers of the China Party, and Doctor Cheng Yunshan, together with the students of the workshop, came to the Fjord Park on foot every day and helped the students solve some technical issues. There were two sponsors of this activity, one was the town committee of the town Tvedestrand and another was the large size ship equipment manufacturer located in the small town. It was unplanned that the branch factory of this manufacturer was situated in Nanjing, of which the products it manufactured were supplied for the Nanjing and Asian markets. Every day, several employees of the town committee came to the site and assisted in the construction actively. In the fjord park, all the five works were featured with specific characteristics.

Entering the park, the first object that came into sight was the tree house built by Chen Yongle, Liu Wei and some others together with Norwegian students. They selected three oak trees as the supporting column for the platform and house about seven meters high from the ground. As every operation was conducted in the open air, there were many unknown elements. The students from both nations took one day in design, where they differed from each other greatly with respect to the design concept. The Chinese students thought to add more supporting columns for the assurance of the stability of the tree house, but the Norwegian students opposed any additions. In mind of the Norwegian students, it was approached mostly to the naturality that the tree house swung with the trees, even grew up with the trees. In the tree house, a bed was designed and purposely built, as the students considered it might be charming to sleep in the tree house while listening to bird's sing. During the process of practical construction, it took three full days in mounting three wooden beams necessary for the platform. It never occurred to the Chinese students that they needed to work on the trees before going to Norway. On the construction site, there was no large size mechanical construction equipment, so they had to make a ladder. In measuring the distance, they had to climb to the top of the ladder and use a rope for the measurement. In order to lift the heavy wooden beam, they specifically made a simplified chain wheel. The climbing was opposite to the tree house, of which the concept was to pile up several three steres cubed in the air, where the platform at each floor protruded in different directions like a dynamic building growing up. Overall, the construction of the tree house went well. Setting up the columns, putting up the platforms and then piling repeatedly, all of these were not technically difficult, but the effects mainly relied on the workload cumulation.

The sloping field near the climbing served as the base for the rotating In and Out. Consideration of the rotating support meant that where a group of rotatable wooden plate walls was built on the comparably sloped field, a playground where the internal space and external space could be interchanged was formed. It could be a place for children where they could combine their own spaces with the rotating wooden plates and discover an interesting phenomenon: through rotating, some external spaces all of a sudden became internal spaces. At the beginning, based on consideration, the realization of the rotation depended on the bearing. But as a result of causes due to processing and maintenance, the relatively simpler double-deck sleeve rotating construction was utilized instead. The playground was located at the sloped field near the rotating In and Out. Consideration of the playground meant that where a group of fluctuating floors were built on the cliffy field along the direction of the sloping filed, children could run, play and have a rest. In aspect of the details in construction, the students from both nations took full consideration upon the demands of the children and the disabled. For example, the spacing of the wooden strips of the floor should neither be too narrow, otherwise children might slip as a result of falling leaves that were frosted in the spacing, nor too wide, otherwise the tyre of a wheel chair might be blocked. As a minimum, both the children and the disabled might pass by safely.

The stargazing was located at the top of the mountain. It was the most special base with the most difficulties in construction. The stargazing allowed for beautiful scenery with a wide view of the field and fascinating landscape. In the beginning, the students considered constructing a landmark building on top of the mountain. But after discussions, they realized that the stargazing was built for the purpose of preventing the expensive astronomical telescope from being wetted and stolen so just a covering might work well. As a result, the design of the stargazing was simplified to building a person high wooden box and a 4m×8m wooden platform. While in design, in consideration of the demands of rest for the observer, a chair for only one person sitting was purposely arranged in the wooden box. As a matter of fact, the wooden box played a role of both covering the telescope and as a big chair with a cap. Skillfully, for meeting the requirements of more observers, this wooden box could move on the platform. When it was crowded, people might move the chair away and surround the telescope for convenient viewing.

Two weeks later, the main construction of the five architectural works had been completed. Wherein, the stargazing, rotating In and Out, climbing and playground had finished construction and put into use. As for the tree house, due to the difficulties in construction and some other technical causes, the upper part had not been completed.

The emplyees of the town committee stated that they would continue the construction until it was finished. At the end of the construction, local children became the first group of users. They played joyfully in the climbing and on the playground and rotated the wooden In and Out curiously. Though Norway is a highly developed modern nation; people here have a strong traditional concept of family. Many families have one or two children. Residents here love children, so it is natural that they give special attention to these large size toys. Frequently, parents put their children on their back, came to the construction site, viewed the construction process and understood the progress of construction. Having completed the construction of the main wooden buildings successfully, students from both nations in the joint workshop were praised and admired by the local town committee greatly. Especially the Chinese students who were unskilled on aspects of the woodworker and woodworking at the beginning, but they studied hard and finally won the reaffirmation and praise of the employees of the town committee. The local residens were thankfull to these diligent and intelligent young men from a nation far away.

Experiencing the building process by themselves, all the students realized the issues relating to the material features, structure pattern, existing form, dimensions, etc. They felt the weight of materials by themselves and understood the significance of the gravity to the architecture. They got to know that just painting and drawing on drafts would never produce an architect. As one of the key building materials, the timbers showed a rich significance with its special material, construction and structural mode. Materials are the physical foundation for any building and the basic issue lies in the rational selection and application of materials. Such a architectural construction teaching course, based on the fundamental features of physics, connection mode of materials as well as the features of construction, etc., through building processes, helped students to consider the logic formed by materials, structure and construction; and comprehend the material and technical base of generating forms.

桥 Bridge

桥的结构原型来源于分布在中国浙南闽北一带的木拱桥。三节苗与五节苗系统通过牛头联系在一起，在荷载作用下相互挤压，共同作用形成一个完整的结构整体，充分发挥了材料的受力性能。

在这里，我们希望从建造的角度探索这种结构类型在构造方式上新的可能。为了方便运输和整体吊装施工，我们用螺栓和铁件连接代替了原有的榫卯和绑扎连接。同时，尝试使用较少的结构拱型结构单元（四榀）与牛头形成的结构框架后再通过斜撑形成稳定的结构系统，不同于传统木拱桥通过并列大量的结构单元以获得稳定性的方法。

The structure prototype of bridge benefits from the traditional wooden arched bridge in south Zhe Jiang and north Fu Jian province,China. Three Jiemiao and five Jiemiao elements are connected by the beams (Niu Tou) which makes them work together as a whole stucture system by pressing each other. The potential of the material's capability of carring load is well exploited.

We tried to look for other construction possibilities of this structure type. For the convenience of transportation and construction, we use steel elements to connect bolt joints instead of the traditional method that used ropes to strengthen the tenon joints. Meanwhile, we tried to use relatively fewer arched structure elements to construct a frame with the beams which was stablized by the diagonal beams. It was also different from the traditional bridge which juxtaposed large numbers of arched elements closely to be stable.

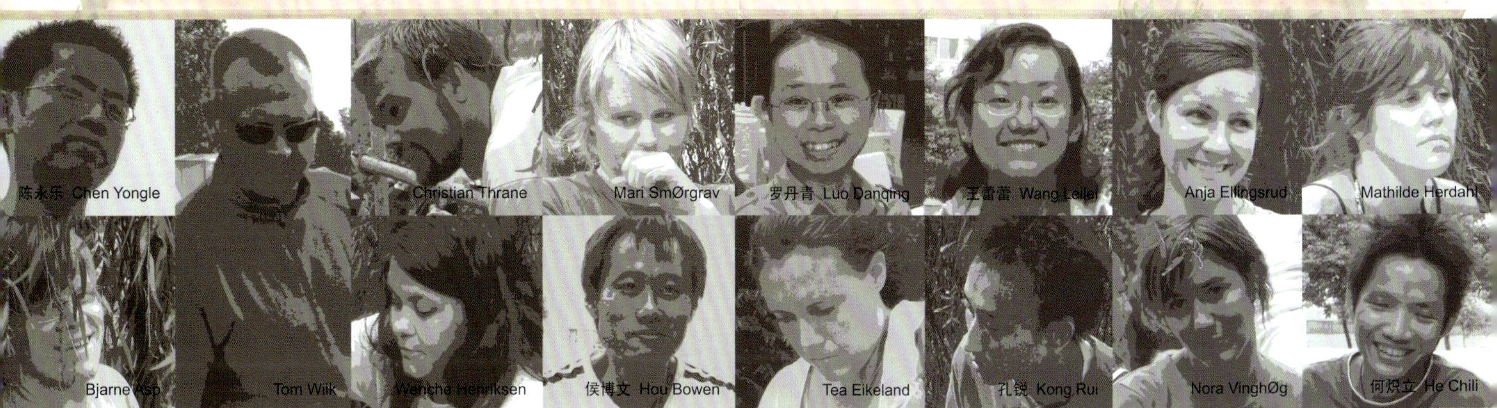

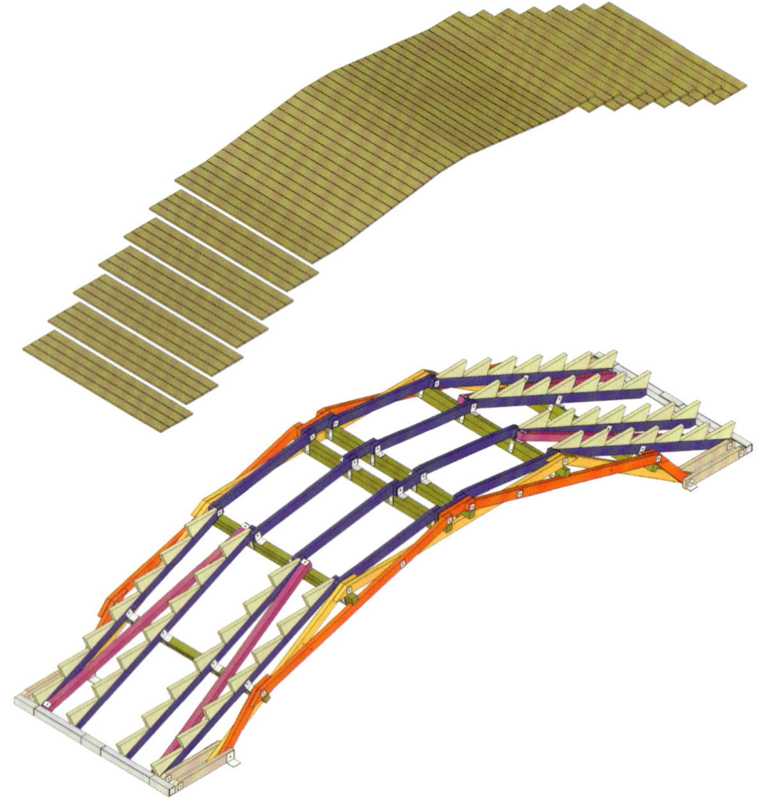

根据设计预制出各木结构构件和金属连接件，并将其组装加工成结构构件的半成品。出于加工难度和成本的考虑，经过多次调整后，金属连接件设计逐渐趋于简单化和标准化。

Prefabricated all the wooden structure elements and steal elements, which were assembled to be structure components of various size. The design of the steal elements was simplized and standardized many times because of the difficulties and cost of manufacturing.

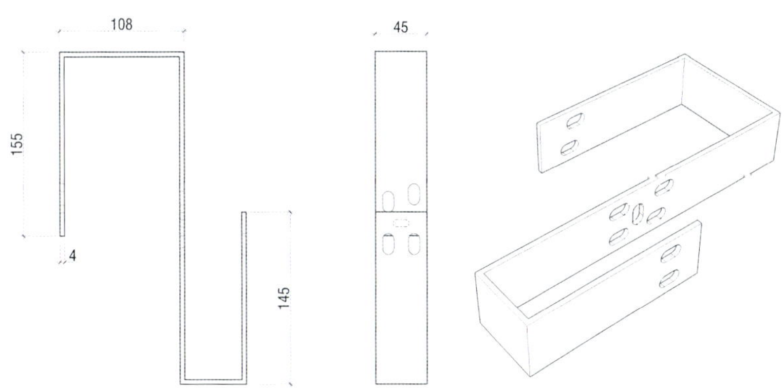
108
155
4
145
45

先在地上放样出１：１平面和结构剖面，根据剖面放样图预制出两榀主结构，然后将其竖立起来，根据平面放样图定位并与牛头、次梁和斜撑连接固定。

Firstly, we drew 1:1 plan and section on the ground, then prefabricated each main structure according to the section. lift them up, brought them to the right position of the plan, and connected each other with the beams.

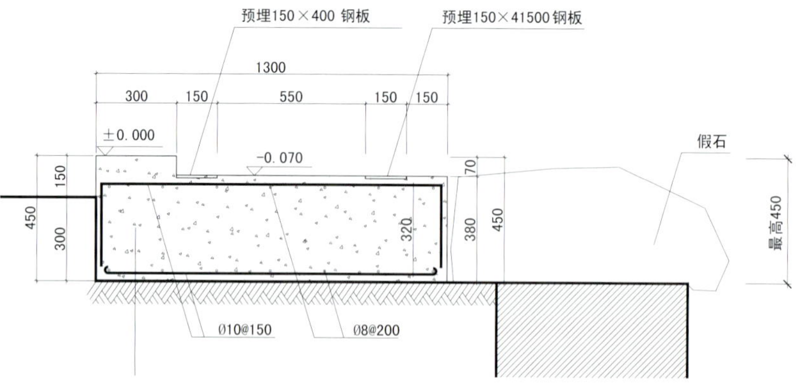

基础大样 Detail of foundation

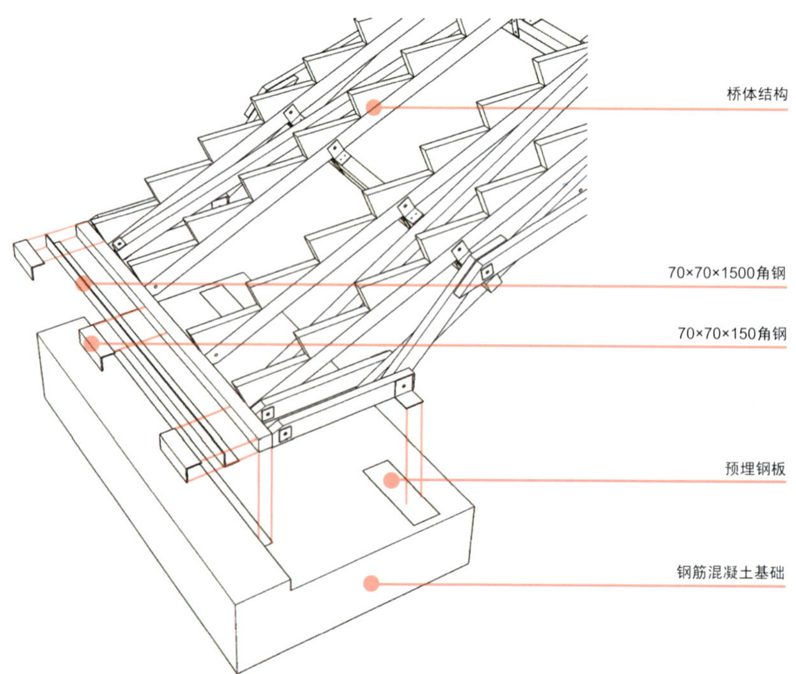

基础构造示意图 Construction of foundation

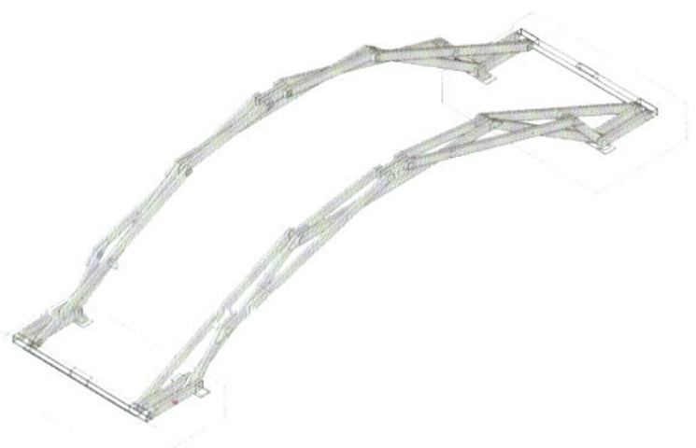

拱的整理与搭建

Construction of the components of arches

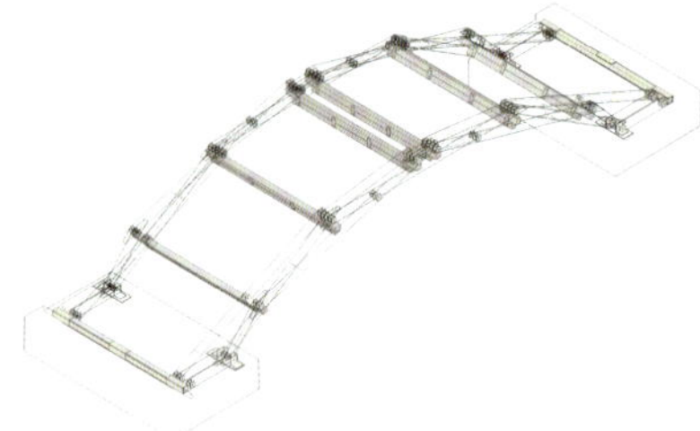

安装牛头

Installation of the main beams

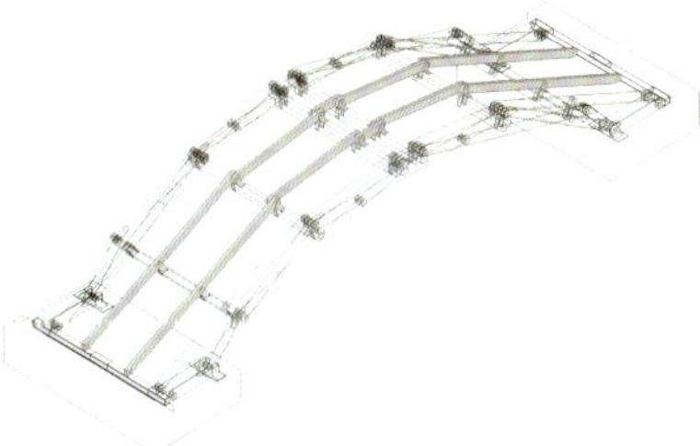

安装次梁

Installation of the secondary beams

IA3

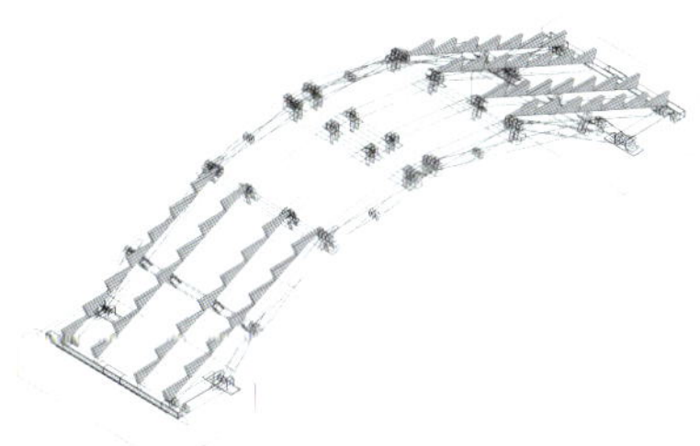

安装三角木
Installation of the “triangle”

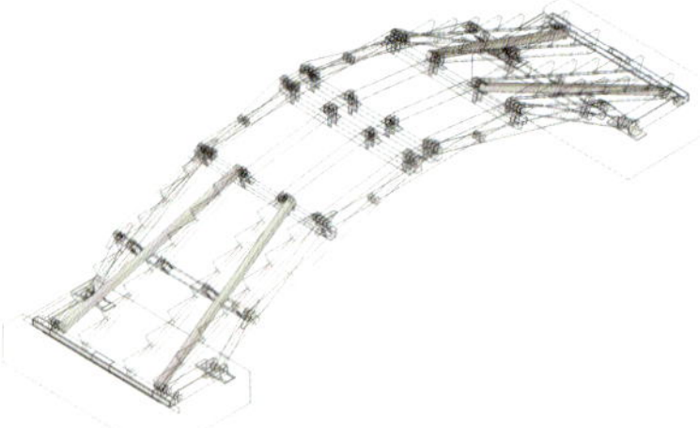

安装斜撑
Installation of the diagonal beams

混凝土基础的浇筑未如预想中的理想和精确，主结构在搬运和重新安装的过程中也产生了误差。校正结构花费了我们不少时间和精力，但我们都认为这是必要和极其重要的。

The construction of the concrete foundation was not precise and perfect enough as we expected. The transportation and reassemblage of the main structure also gave birth to some errors. We all considered the rectification to be necessary and essential although we spent a lot time on it.

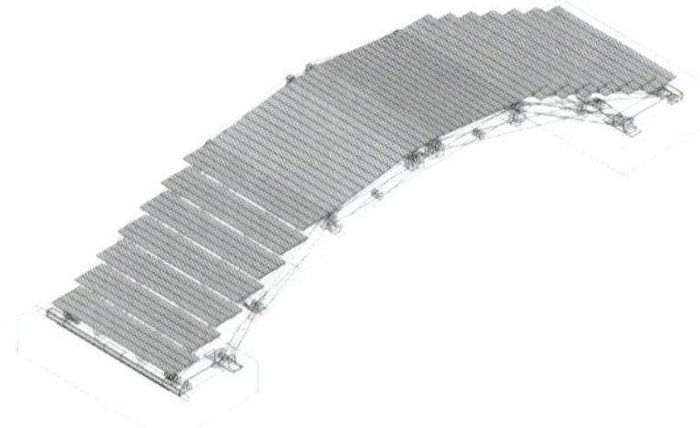

铺设面板
Paving the steps

亭作为供停留之场所的空间形式，是与桥联系的动线的端点。事实上，对建造设计的关注远远超出了结构物本身。一方面，这很大程度上决定了结构物的最终形态，例如出于对可加工材料尺寸的考虑，我们采用了对角线方向为主的结构框架。另一方面，它也激发出了一些有趣的设计结果，例如座椅的设计。它源于功能需求，发展于建造试验，最后却回到了结构物本身，成为整合结构整体性的关键部件。在这里，我们找到了设计的支点。

Serving as the place for staying, pavilion could be considered the terminus of the moving from the bridge. Actually, we impose much more emphasis on the construction process than the structure itself. On one hand, it resulted in the final form of the structure to a great extent. For instance, we selected the two diagonals of a square as the main directions to make our main frame according to the size of the offered timber. On the other hand, it also inspired some interesting design as well, such as the design of the seat. It was brought forward for the needs of the users, developed in the experiments of construction, and became one part of the structure finally which was an important element for the integration of the whole system. Here, we had found fulcrum of design.

屋顶是建造的重点和难点，我们先在室内对其结构进行了准建造以研究真实建造中可能出现的问题。将其移建到室外，还有一些其他的问题要考虑，比如将原先600mm的短柱替换成2550mm的长柱，并考虑其与基础的连接方式。为此，我们设计了相应柱础铁件和屋顶的覆盖形式。

The construction of roof was the most important and difficult part the whole project. We constructed the roof structure indoors first to study the possible problems of future construction. To build it onto the site, we had to think about something else such as the connection method of the base with the pillars which would be replaced with the 2550mm ones for the short ones of 600mm. Therefore, we designed a special steel elements for that and the form of the shelters.

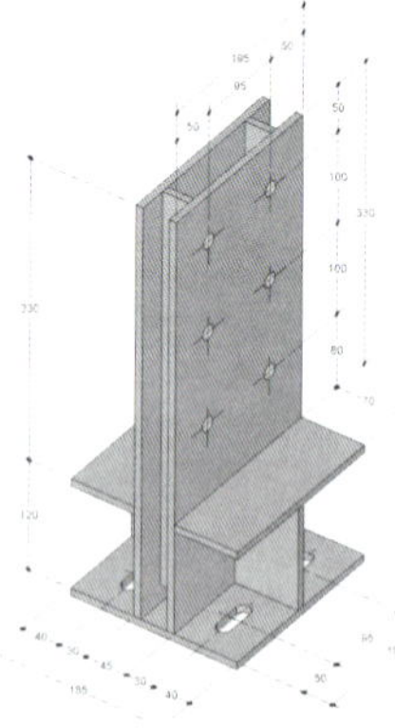

柱础铁件大样

Detail of column base

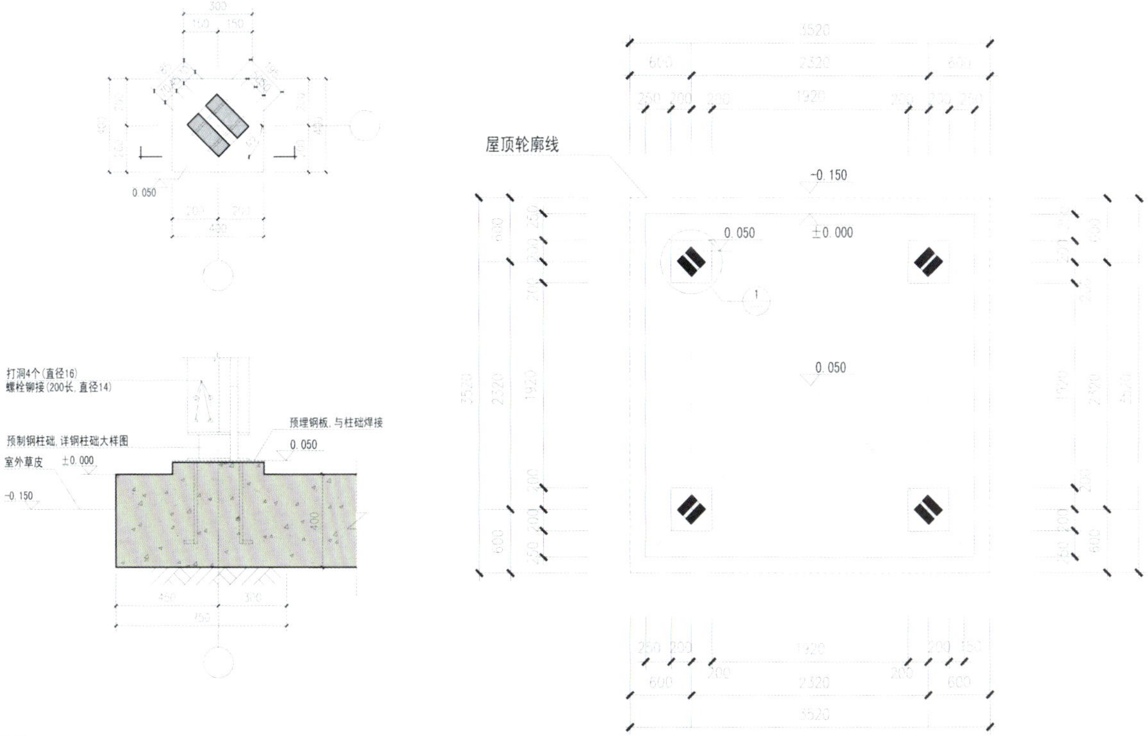

打洞4个(直径16)
螺栓铆接(200长, 直径14)
预埋钢板, 与柱础焊接
预制钢柱础, 详钢柱础大样图
0.050
室外草皮
±0.000
-0.150

屋顶轮廓线
-0.150
0.050
±0.000
0.050

基础平面图
Plan of foundation

剖面图 Section

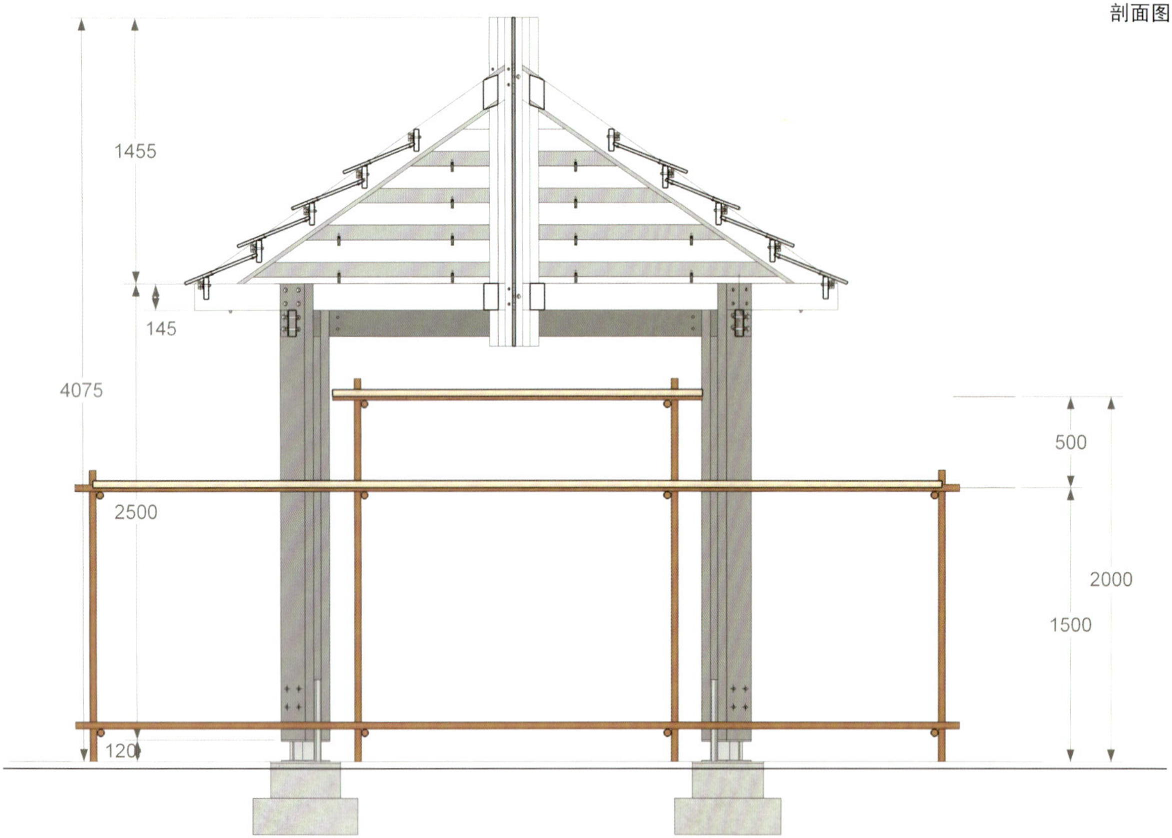

1455
145
4075
2500
120
500
2000
1500

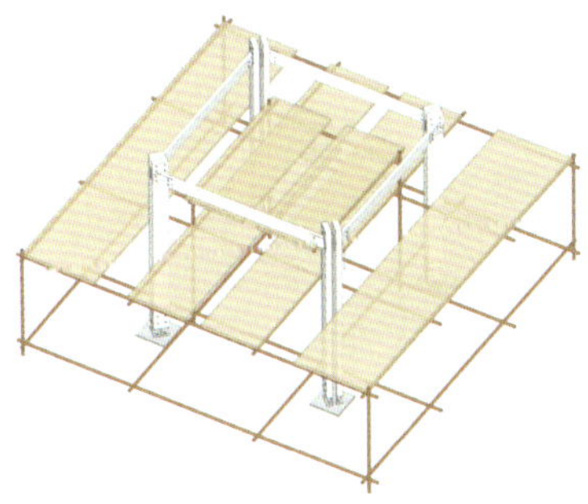

立柱及圈梁安装
Construction of the columns and beams

安装边梁，校正调整使其围合成一个独立稳定的构架之后，将柱础与基础上的预埋钢板焊接固定。

Welded the columns to the foundation to stablize the structure, after rectifying and fixing the framework to the right position.

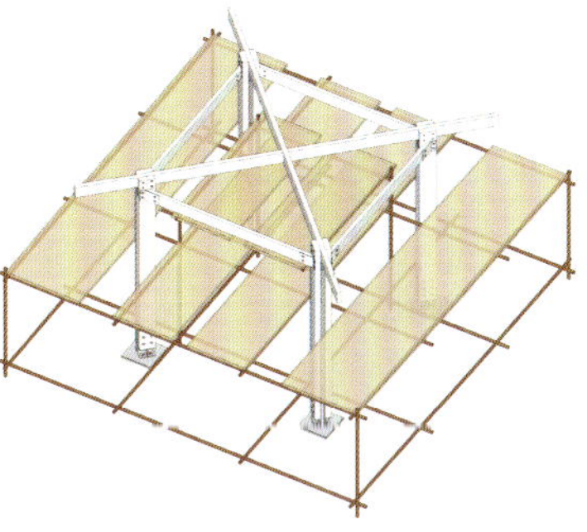

安装十字梁
Construction of the crossing beams

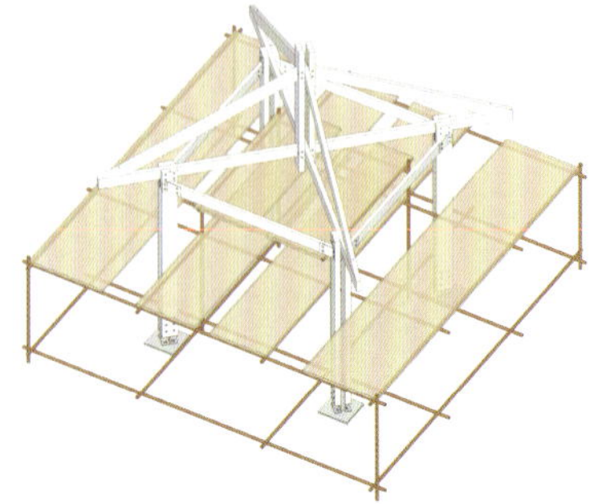

安装雷公柱和斜梁

Construction of central post and diagonal beams

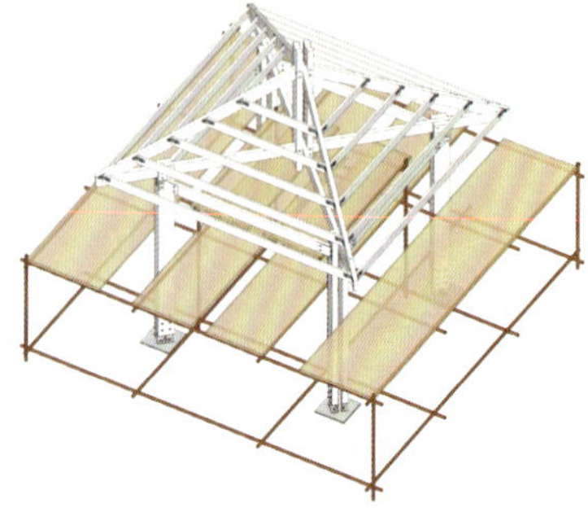

安装横撑
Construction of the purlines

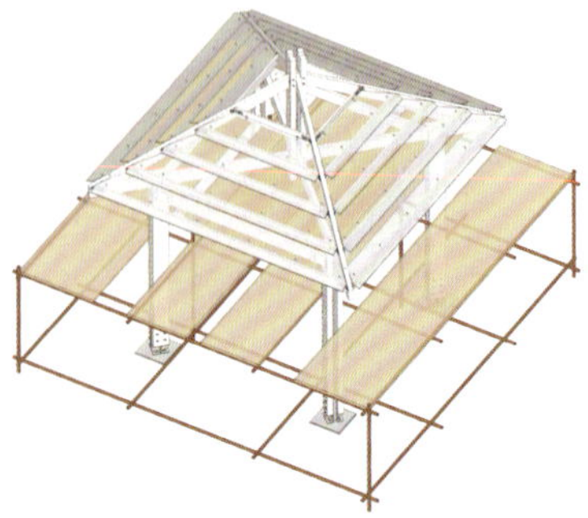

安装阳光板

Installation of the translucent slabs

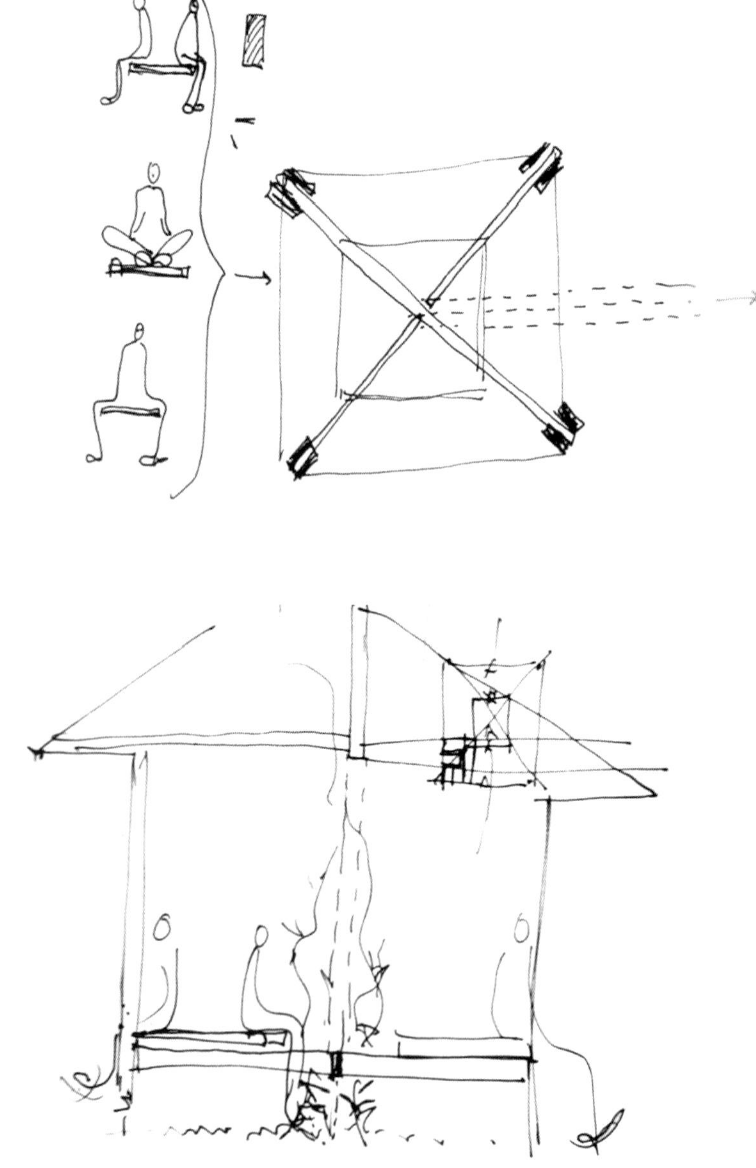

为亭子加上座椅是在建造过程中，和挪威同学讨论后新增的设计。我们想要一种不同于传统的（如美人靠、桌凳之类）形式，并在现场进行了构造方法的试验。实施方案是在 800mm 的高度上，采用与屋顶结构相同的构造方式，并以钢索与顶部系统联系，表面铺上木板。在这个台面上，使用者可以坐卧站立，尽其可能地活动。更重要的是，它将四根柱子联系在一起，与顶部的圈梁共同作用，大大加强了结构的整体性。

The design of the seat was the result of the discussion with the Norwegian students during construction. We wanted some new form different from traditional ones, and did some experiments of possible construction manners at the site. It turned out to be at the height of 800mm which was applied the same structure form as the top, and clung to the roof structure with wires. With wooden plates on it, users could sit ,lie down or stand as they like. Moreover, it served as a hoop to hold the four columns together, and strengthened the whole structure.

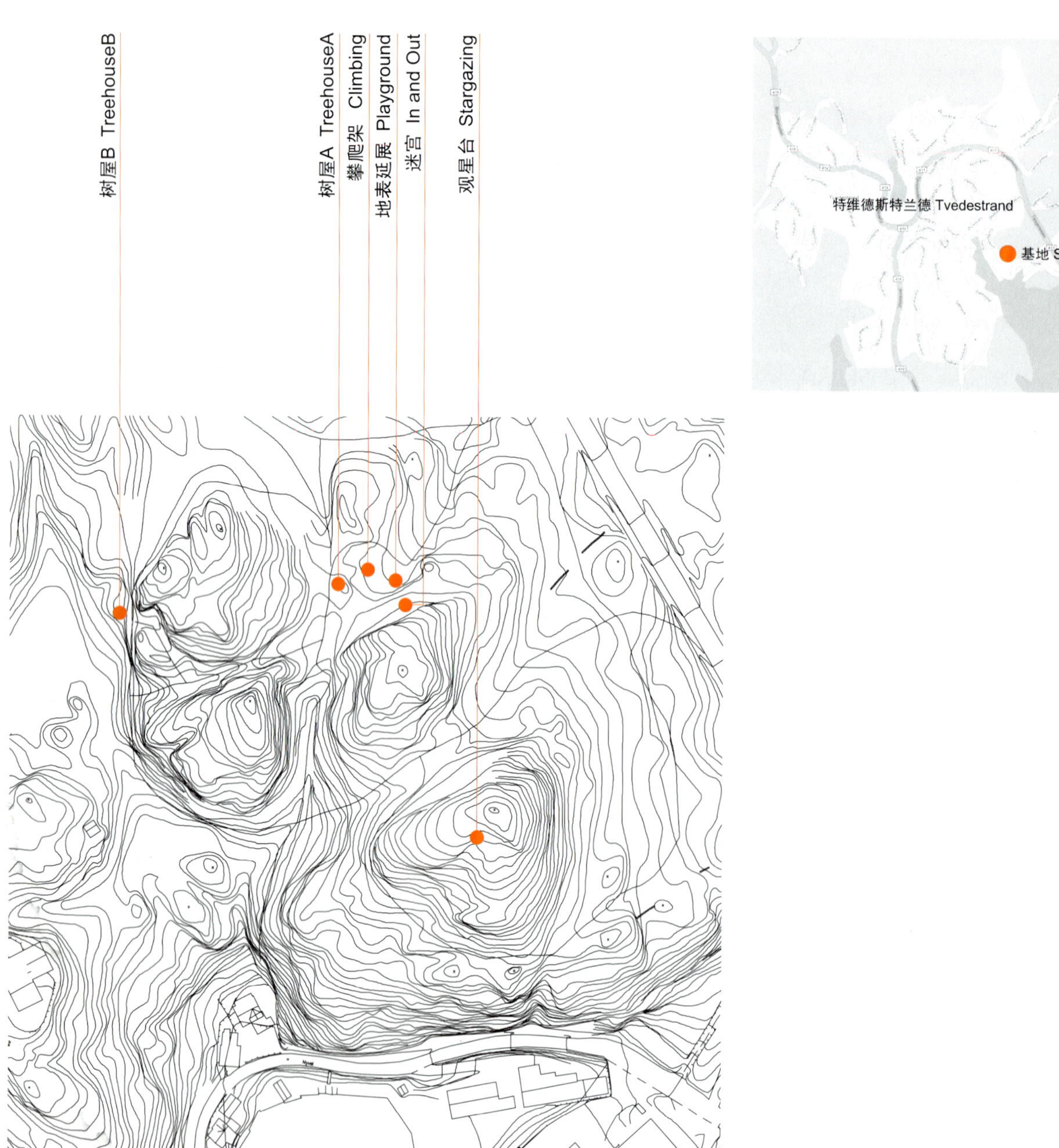

树屋B TreehouseB
树屋A TreehouseA
攀爬架 Climbing
地表延展 Playground
迷宫 In and Out
观星台 Stargazing
特维德斯特兰德 Tvedestrand
基地 Site

特维德斯特兰德
Tvedestrand
2006.10-2006.11

2006年秋　特维德斯特兰德，挪威南部
南京大学和奥斯陆建筑与设计学院联合工作营

星期一，2007. 5. 7写于奥斯陆
沙米・林塔拉
建筑师
挪威国立理工大学教授 II
奥斯陆建筑与设计学院教师

当我在筹备这个工作营时，我的计划是基于之前类似的国际工作营的经验的，一个是 2005 年谢菲尔德大学和特隆德海姆挪威国立理工大学，另一个是 2006 年东京艺术大学和特隆德海姆挪威国立理工大学。它们都是在同一个森林公园里进行的。

这次我们拥有一个富有成效的起点，在南京由赵辰教授领导的和我作为访问教师参加的在奥斯陆由尼伯（Kolbjørn Nesje Nybo）先生领导的教学活动都对木建构拥有共同的兴趣。

我与挪威南部特维德斯特兰德一个适合举办工作营的客户保持有联系，他们是一群想要以非营利的方式丰富当地公共与文化服务的年轻人。

由他们提出并且正在实施中的一个大型项目是对邻近森林中的一个公共公园的开发，它对肢残人士、老人和小孩之类的人群也是开放和安全的。他们得到了公共基金和当地政府规划上的支持，允许进行短期的建造活动。

同时，这个项目将以实验性项目的性质得到两个主要的政府基金组织——挪威创意基金（Innovasjon Norge）和住房储备基金（Husbanken）——的资金支持。通过这种活动，他们希望能够找到设计这种全天候、对所有使用人群都开放的室外场地的创造性解决方案。

然而对于我们来说，最重要的是这个项目能为学生提供什么样的学习平台。为此，我试图在工作营中包含几个主题：

1. 实时设计

有限的工作营时间不允许我们对设计作过多的推敲，而需要按草图的设计开始建造，并随时根据实际的建造对计划作出调整。我希望这个过程能帮助在草图设计中的理想情况与设计意图和经济地使用现场材料与分配工作时间之间找到一种平衡。

2. 与客户的对话

计划必须与客户充分沟通以体现他们的愿望和喜好。同时，客户也会在建造现场观察材料和工作时间的使用情况。

如何向非专业人士表达概念并在最终成果中保持设计中的关键内容也是十分重要的。

3. 亲手建造

每一个建筑师对建造材料质感的亲身体验是很有好处的，诸如重量、刚度、节点。另外，学会一些基本的建造逻辑和手工技巧对于将来与真正的建造工人进行沟通也是十分有用的。

4. 跨文化合作

在了解其他人的设计和工作方法的同时，学生们正在建立一种在未来的职业生涯中可能会使用到的国际网络。

大概在工作营结束了一个月之后，我们和奥斯陆的学生们视察了一下项目的情况。对于每一个人这都是一次成功的旅行，重新将他们的工作作为一处功能性的建筑进行反思，而不仅仅是一个有感情寄托的物质性的成就。

看到我们设计的结构物能独立地继续他们的生命是十分有益的。从现在开始，他们被感知只是其固有的物质存在，而不需要任何哲学性的解释。

客户对于结果是满意的。当我写下这段文字时，那些建构物很可能正在被公园里成群的孩子们使用着。公园将进一步地开发而我们的工作也将作为后续基金使用方式的参考对象。

从这点来看，我们也许可以认为我们在有限的时间里出色地完成了工作。然而，我最希望学生们能从这次的工作营中学到些什么，以及我们奥斯陆能和南京大学继续已有的成功合作。

TVEDESTRAND WORKSHOP

In Oslo on Monday 7.5. 2007
Sami Rintala
Architect
Professor II NTNU
Teacher AHO

Autumn 2006 Tvedestrand, South Norway
Cooperation workshop between Nanjing University and AHO Oslo

When I was preparing this workshop I based the planning on experiences from previous similar international workshops, one with Sheffield University and NTNU Trondheim in 2005, and another with Tokyo Geidai University and NTNU Trondheim in 2006. These both took place in this same forest park.

This time we had a very fruitful starting point, the shared common interest in wooden structures both in Nanjing class under Professor Zhao Chen and in Oslo class lead by Mr. z, which I was participating as a visiting teacher.

I had a suitable contact for a workshop client in Tvedestrand South Norway, a group of young people who want to enrich their local public and cultural services on non-profit bases.

One large task they have invented and are currently working with is to develop further a public park in a nearby forest area, accessible and safe for wheelchair users, elderly and children alike. The client had received public funding for the project as well as planning support from the local authorities, guaranteeing building permissions in short time.

At the same time, this project would be followed up as a pilot project by Innovasjon Norge and Husbanken, two major state funding organizations. Supporting this kind of activity, they are hoping to see creative solutions when it comes to designing accessible structures for all kind of user groups in this kind of all-year–round outdoor situation.

The most important issue for us, however, is what kind of learning platform for the students are we offering with the project. Considering this, I tried to include the following topics in the workshop:

1. Design online

The short timeframe of the workshop did not allow too much time for reflection over the design, this time it was necessary to start constructing with sketches, and changing the plan while learning from the real structure. I hoped this process would help reaching the balance between the idealism and intensions of the design sketch and the economy in use of material and work time at the site.

2. Dialogue with the client

The plans had to be discussed with the client to include their wishes and preferences. At the same time, the client was present at the construction site surveying the use of material and work time.

It is important to learn how to convey an idea to a person outside professional circle, and how to keep the essential parts of the design present in the end result.

3. Constructing hands on

It is good for every architect to have a bodily experience of the qualities of construction materials; weight, resistance, joining. Further on, it is useful to learn some basic understanding of construction logistics and manual skills so that one may communicate with credibility with the actual builders later on.

4. Cross-cultural cooperation

While getting to know each others' ways of designing and working, the students are creating an international network they may use in future in their professional lives.

After one month or so, when the dust had fallen down again, we visited the park with the Oslo students to have a review of the projects and to see how they stand time. This was a very necessary trip for all to reflect upon their work as a functional piece of architecture, not just a physical achievement one is emotionally attached.

It is very healthy to see that the structures we design will continue their lives independently. From now on, they can only be perceived as they physically are, without any philosophical explanations.

The client has been satisfied with the results. The constructions are probably being tested by groups of children in the park as I write this. The developing of the park will continue further and our works are used as a reference for applying future funds.

Thinking of this only, we may consider we have done a good job in a very limited time. However, I mostly wish the students have learned something useful during the workshop, and finally that we here in Oslo are able to continue the successful cooperation with Nanjing University.

建造的初衷是为孩子们提供攀爬、翻滚甚至跌倒的场所。提供多样的空间成为设计的出发点。折板不仅确定了“板上”的空间，同时，通过和坡地对话，提供了“板下”空间。不同斜度折板的延续，若干行地板的组合，提供了丰富空间的可能。

基于对建造和无障碍设计的考虑，我们决定由平行的若干地板平行拼合成大“地板”。由于建造时间有限，决定只用三组地板拼合，但提供加建的可能，使之更接近“地板”，为孩子们提供更大的活动平台。

The original intention of this construction was to provide the children a space to climb, roll and even slip. The diversity of the space was the concept of this design. Table-flap defined not only the space of “up table-flap”, but the space of “below table-flap”. The continuity of table-flaps which differed in angle, and the assemblage of several row provided the possibility of various space.
Considering the convenience for construction and the use of the disabled, we decided to assemble several parallel rows. The limited time made us to build three rows at first, but provided the possibilities for future construction of more playground for the children.

Bjarne Asp 罗丹青 Luo Danqing Jan Gunnar Nora Vinghøg Tom Avger 王蕾蕾 Wang Leilei

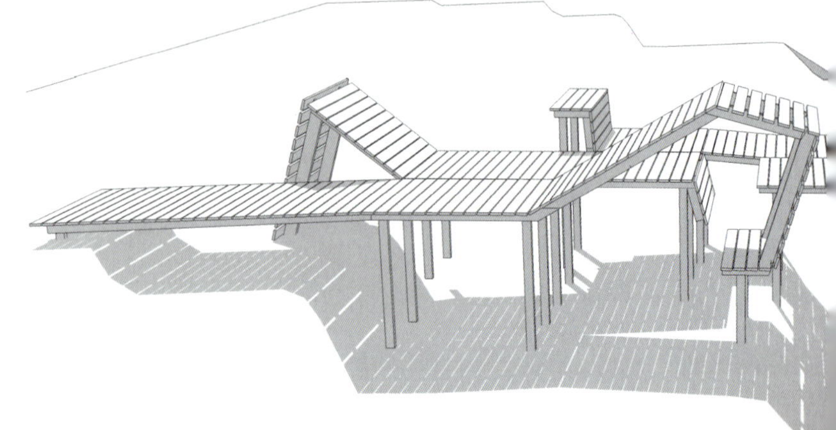

在场地上用绳子确定三行地板的位置。根据设计的形态确定基础点。由于在需要埋设基础的地方，地基的情况并不都是岩石，而如果浇筑大体积的混凝土，时间不允许。于是我们在需要的基础处挖坑，填以岩石，在岩石上钻孔。

We used ropes to define the locations of the three rows.The points to lay foundations were determined by the designed form. As not all the places had rocks underground, and it would take too much time to cast concrete, we dug holes at the wanted places, stuffed rocks, and drilled holes in them.

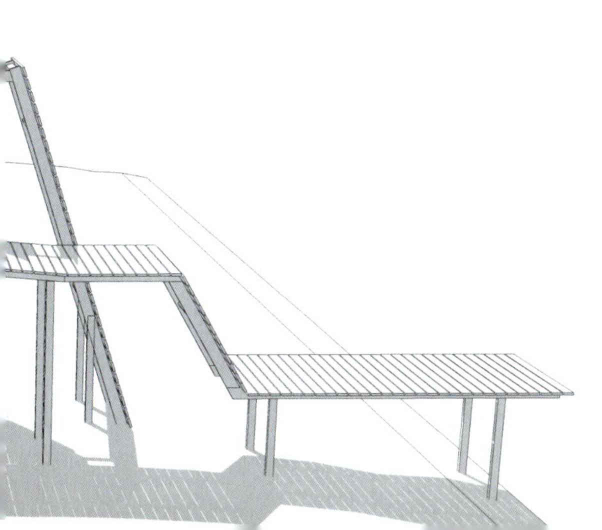

结合现场的树木和坡地设计每行地板的起伏，并考虑为轮椅提供比较平缓的走道。折板的斜度主要靠现场感受来确定。考虑行走和攀爬两种活动，面板的间距有两个尺寸，45mm 提供攀爬时手脚放入的空间，15mm 提供轮椅通过的尺寸，这个尺寸不能过小，否则树叶容易夹在缝中，雪天膨胀对木板不利。

适合攀爬活动 for climbing

适合轮椅活动 for wheelchairs

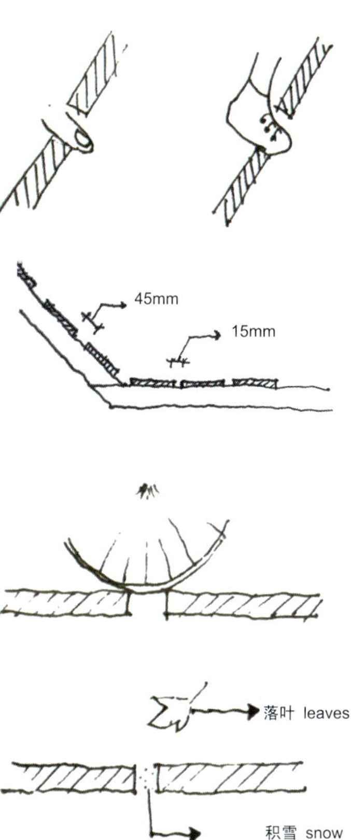

Took the trees and the sloping site into consideration to design the wave of each row, and provided walkways for wheelchairs.The angles of table-flap were decided by the situations at site.Walking and climbing required two dimensions of space between boards:45mm for climbing and 15mm for wheelchairs. The latter must not be too small. Otherwise, the leaves would be easy to stuck in the gaps, and this was disadvantageous for the boards at snowy days.

有趣的是，工程尚未完工，就吸引来不少孩子在上面玩耍。这也增添了我们对它未来的信心。

Interestingly, many kids had come to play on it when the project was still under construction. It made us more confident for its future.

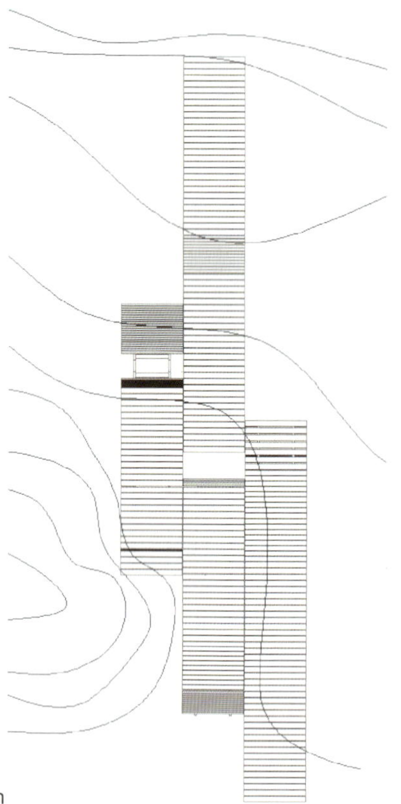

平面图 Plan

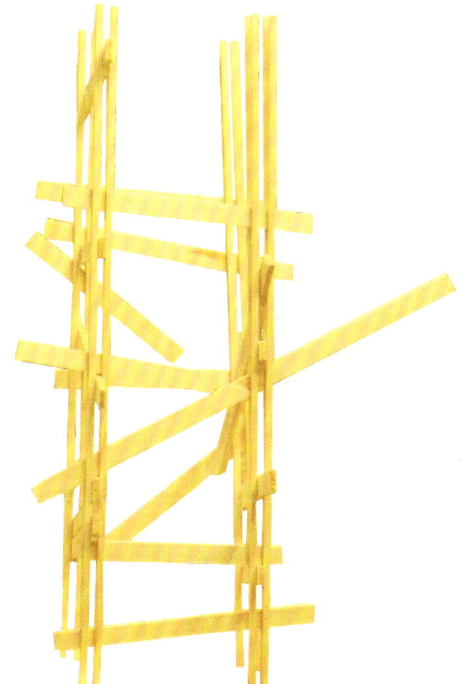

攀爬架小组工作的目的是在林间为儿童和残障人士建造可攀爬设施。

设计最初是从对基地环境的认识开始的：橡树环绕的几米见方的空地，以整石为底。这样的环境促使小组成员们去思考、去怎样创造设施来帮助人们获得与环境结合的攀爬体验。

整个结构也以“层”作为竖直方向上的划分单元。“层”不仅作为结构的竖向单位，也标示着攀爬活动的阶段逐“层”攀爬，人们可以清晰地感受到树林中空间的竖向变化，并最终如鸟儿一般置身树冠中。攀爬架小组仿佛并不是在构筑一个物体，而是在设计一系列密林中的人们攀爬和休憩的路径。

设计与建造同时进行，期间我们基于一种原则不断地修正设计，即构建尽可能简单的木框架来实现引发和满足使用者的多种攀爬方式。

The work of the climbing team aimed to construct installations to be climbed by children and disabled people in the forest.

The initial concept of design came from the understanding of the environment: An oak stood upright in an open space which based on an entire stone. Such environment prompted team members to think about how to create installations which could help people to gain the climbing experience and enjoy the surroundings. ‘Layer’ was designed as the vertical demarcation unit of the entire structure, which also indicated the different phases of the climbing. Climbing layer by layer, one could clearly feel the vertical space changing in the forest and reached the crown of tree like the birds finally. It seemed that we were not constructing installations but designing the route which people could climb and relax.

Design went forward with the construction. We constantly revised the design basing with such principles: using the simplest wooden framework to construct, while meeting the need of the users and creating a variety of routes to climb.

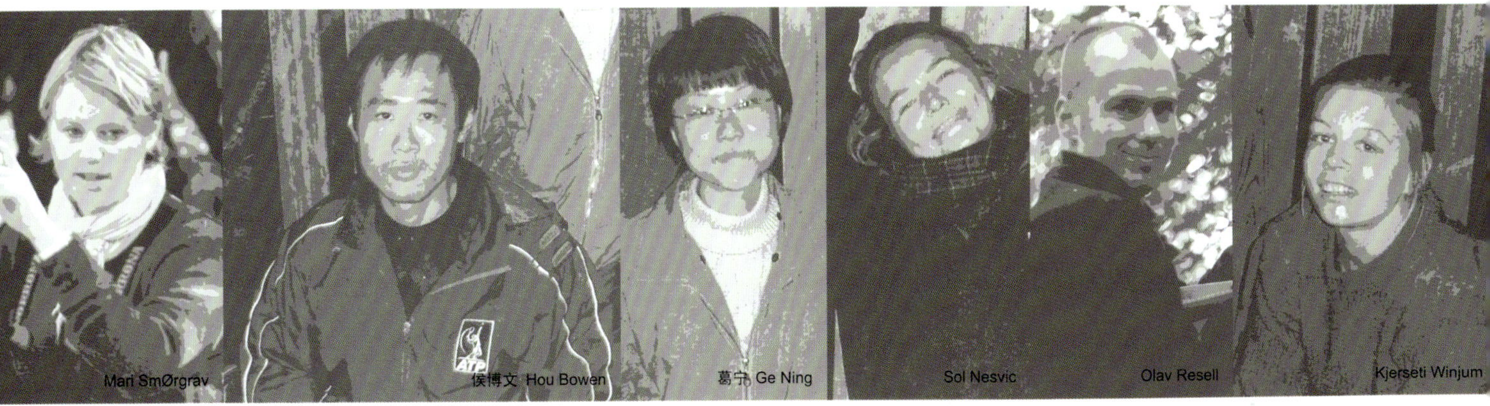

确切的建造位置是组员们在基地上“框选”出的，此处位置和尺寸上的不可精确描述的特性使得整个建造活动的每一步都必须参考和应对上一步建造形成的条件，而不是按照一份几天前就确定的图纸实施。

The exact place to construct was located by placing a frame. The location and size was hard to accurately defined so that every step of the entire construction activities must base on the former construction, instead of constructing according to the drawings faithfully which was determined a few days ago.

小组将标准截面的木料从中一分为二,四根组合成为柱子。这样不仅可以将用于竖向连接的节点巧妙地结合其中，而且十字形的中缝可以在两个方向自由组合放置横向杆件，来设置不同的攀爬方式和难度。同时，既有利于柱子在雨雪后的快速干燥，也可以避免单根木料的单向弯曲。

The team members divided the standard wood into two pieces. The column was made up of four pieces so that joints for vertical connection could be installed into the column. Furthermore, the cross-shaped column could support horizontal bars in two directions which created different ways of climbing and diverse difficulties. It was good for the wood to dry as soon as possible and avoid one-way bending as well.

现场的测量与调整司空见惯，新的问题层出不穷，这些刺激我们对原有的设想不断地修改。

We frequently measured and adjusted to new conditions at site, as new problems kept driving us to modify the original concept time and time again.

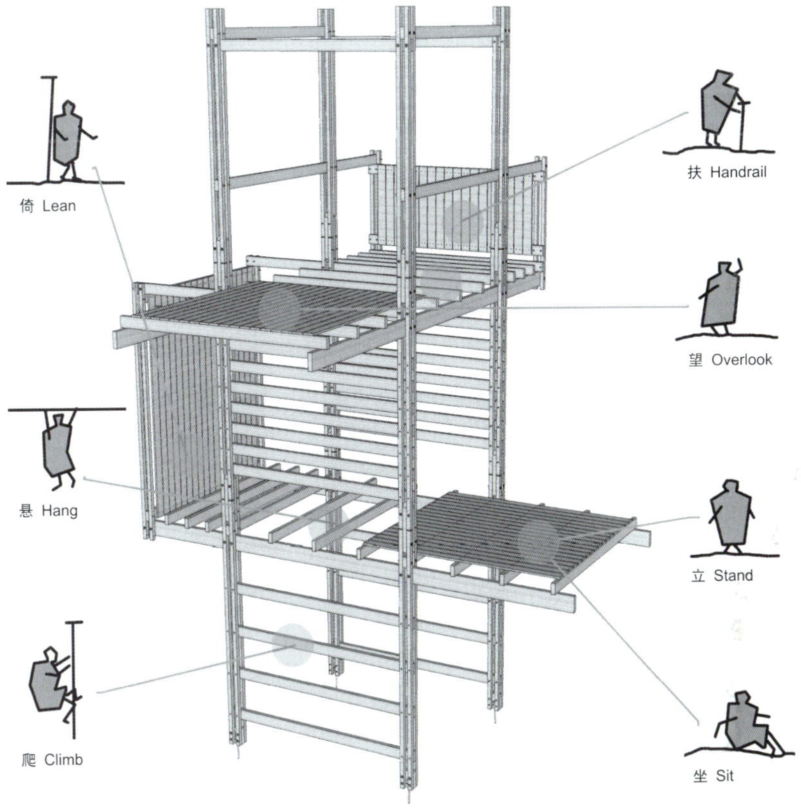
倚 Lean
扶 Handrail
望 Overlook
悬 Hang
立 Stand
爬 Climb
坐 Sit

纵观整个建造过程，攀爬架从地面开始生长，因环境而成形，最终被茂密的枝杈穿梭其中融入林间，正是这个建造过程的诗意所在。

Throughout the entire construction process, the 'climbing' started from the ground, shaped by the surroundings, and became part of the forest at last. What a poetic construction!

由于界限的界定才会产生内与外的区分，同样，如果界限模糊就会带来内与外的不确定，在这里我们希望给孩子提供一种不同于建筑实体的构造物，它让玩耍于其中的小孩通过身体的移动体会到界限的微妙所在，同时构筑物在优美的树林中也消解了自身和环境的界限。

现实中的迷宫为我们概念的实施提供了一个很好的平台，但是对于环境的尊重迫使我们消解迷宫自身所包含的强烈的自我姿态。我们在场地中不规则地设置转板，其位置的确定是来自于场地中的岩石和树木的方位。这一方面体现了对于环境的思考，另一方面转板在实现了界限概念的同时也发展出了迷宫的形式。

The division of the inside and outside derive from the boundary, that is to said, the unsureness of the In and out derived from the ambiguity of the boundary. Here, we wanted to provide a construction to the children that was different from buildings, in which the kids could experiment the subtle existence of the boundary by playing. At the same time, the construction dissolved itself in the graceful woods.

The labyrinth provides a platform to conduct our idea. Yet, we had to dissolve the strong characteristic of labyrinth, making it become part of the environment with respect of the site. We designed to place many rotating broads in the site unregularly, whose locations depended on the situation of the rocks and the trees in the site. On one hand, they could fit the site well. On the other hand, the rotating boards achieved the conception of boundary while developed the form of the labyrinth simultaneously.

彭嫱 Peng Qiang　Martin Stockinger　Mathilde Herdahl　蔡伟森 Cai Weisen

通过拉线来确定转板的统一标高。再根据场地中的岩石确定转轴的平面位置，没有岩石的地方通过浇筑混凝土来固定。

Figuring out the height at which the boards would be by running a thread. Located the pines by the situation of the rocks in the site, and concreted a foundation in case that there was no rock.

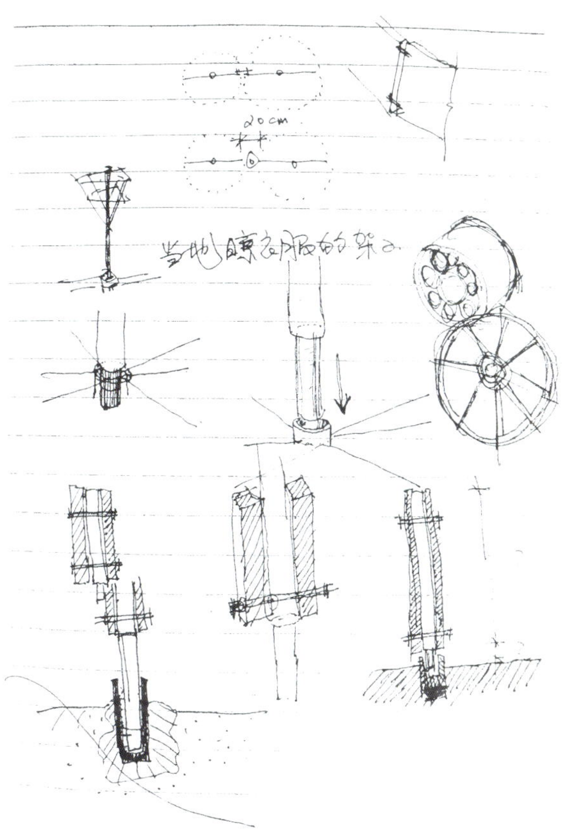

转轴是最关键的部件，它的设计借鉴了当地一种特有的晾衣架，并使用不锈钢制造。在木板上安装不同质感的材料：穿孔板、沙网、木条，让板面在转动的同时也带来触感上的差别。

The pine was the key elements whose design refered to a special kind of local rack, and was made of stainless steel. Putting various of materials on the boards, such as perforated plates, wire netting, woods, could give birth to some special tactility while rotating them.

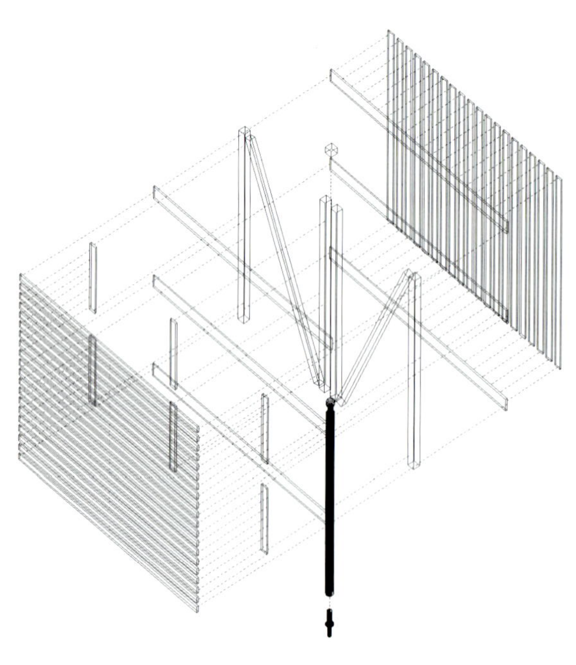

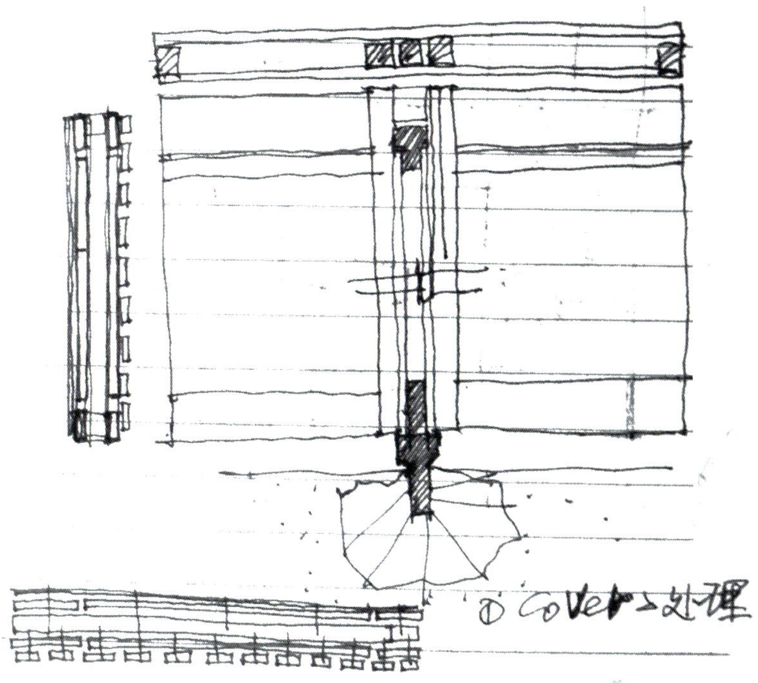

树屋A TreehouseA

客户的设想是能在未来的几年时间内在公园里构建一个树屋群供周围的居民使用。因此，我们不仅要搭建一间树屋作为整个建设计划的基础和起点，更重要的是对未来发展的可能性提出可行性建议。客户对于树屋的要求是：一个温暖舒适的地方，并能放下一张桌子，两张床。我们决定选择一块小高地上的三棵橡树作为主要的结构支撑，这是平衡树屋尺度、安全性和客户要求的结果，因为他们不希望树屋有外加的独立支撑而是和树木一起“生长”。地形的高差也有利于降低施工的难度。三角形的平面是顺应三棵橡树自然围合出的空间的结果。

The clients wanted to construct a group of treehouses for the residents in the coming years. Therefore, we were not only going to build a treehouse as the foundation and threshold of the whole programme, but also to provide the client the praticle possibilities for the future development, which was even more important. The clients wanted the treehouse to be a warm and comfortable place where they could place one table and two beds. We decided to choose three big oaks on a small plateau as the main structure by balancing the consideration for the proper scale and the safety of the treehouse with the requirements of the clients, that they didn't want the treehouse to have any extra structure but 'grow up' together with the trees. It was also easier to construct on a relatively higer level. The triangle plan is out of the natural space formed by the three oaks.

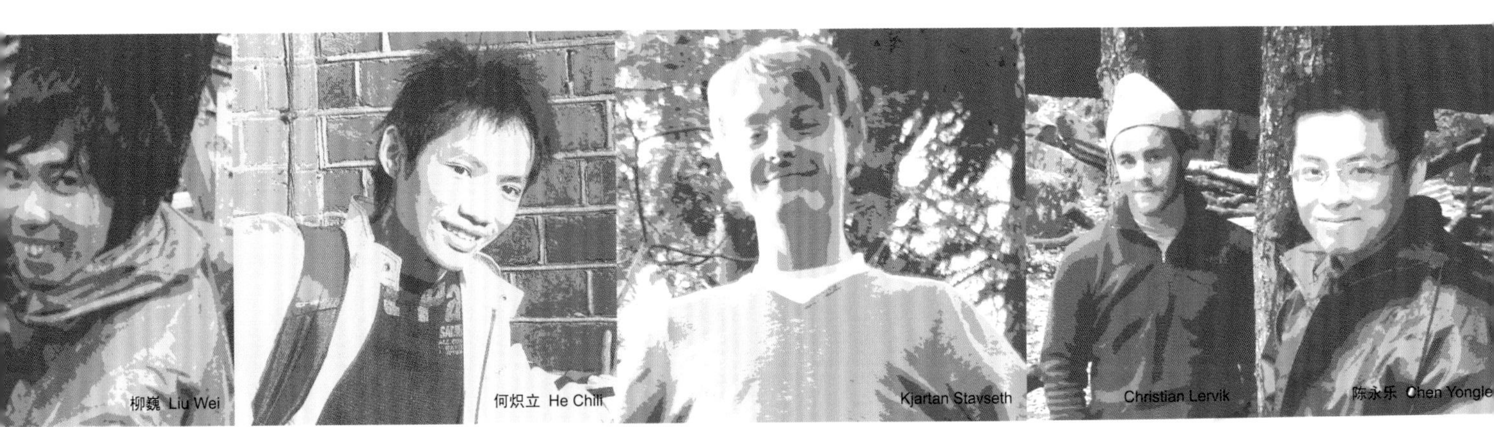
柳巍 Liu Wei　何炽立 He Chili　Kjartan Stavseth　Christian Lervik　陈永乐 Chen Yongle

在 5m 的高度上工作对我们是个很大的挑战。树木的自然生长给设计和施工带来的困难首先是尺寸测量和空间定位上的，其次是建造上的。木料本身成了测量的工具，现场定位，现场测量，现场制作，现场安装。我们利用剩余木料制作简易的爬梯，绳子和树木的枝杈作为吊装木梁的简易滑轮，周边的树木也被我们用来固定绳索。随着建造的进行，现场的材料和不断变化的条件不断刺激着我们修改设计，寻找最可行的建造方法。

Working on a high level of five metres was a big challenge. The natural growth of the oaks raise the difficulties for measuring, fixing and construction. The timber themselves became the rulers that we lifted them to a certain level, measured and marked immediately. And then we cut and fixed them. We took the rest timber as the material of ladders, ropes and branches as a simple chain wheel to elevate the wooden beams, and the trees around became the stakes where we fastened the ropes. As the construction going on, every material we had in hand and the changing conditions continuously stimulated us to change our design, and figure out the most practical solutions.

客户对于结果是满意的。虽然我们没有完全完成整个结构的搭建，但是他们已经能够看到未来发展的前景。挪威的朋友告诉我们，他们将在暑假回到这里将其完工。

The clients was satisfied with the work. Although we didn't completely finish the whole treehouse, they could see the prospect from it. Our Norwegian friends told us, they would returned at summer vacation to finish it.

"Opphengt" / "hanging

prefabrikerte hytler i heises opp i treet.

在有限的时间、预算和劳动力的限制条件下，我们根据现场环境提出了我们的概念：一个“可移动”的树屋。概念的由来是基于用非传统的方式建造树屋的想法，它将挑战它的使用者。客户也对这个能随风摇摆的概念颇感兴趣。

我们最初设计的树屋像一个灯笼。它很小，平面大约为 1.5m×1.5m，高 2.5m，只能容一人栖身。很遗憾，我们现有的资源无法完成这样复杂的建造。所以我们简化了设计，三角形成为新树屋的主题，可以容納两个人。

With limited time, funds and capacity for work, we decided to pursue a concept: a "flexible" tree hut, which in theory, seemed in agreement with the given conditions.. The concept was based on the idea to make spectacular huts in a non-traditional way that would challenge the users. The clients were also interested in the 'hanging around' concept.

The tree house we originally planned to build, were to symbolize a lantern. It was to supposed to be rather small, with capacity for only one person. Approximately 1.5m × 1.5m in plan and 2.5m in height. Unfortunately this construction was too complicated with our resources, and we had to simplify the original design. We based our "new" tree hut on the triangle, and made it for two persons instead.

Tina Bjorneset Tea Eikeland Anja Ellingsrud

预制多种尺寸的可以适合不同树木和地点的树屋，重点考虑其灵活性和适用性（包括残疾人）。

Prefabricated tree huts in various sizes, specially adapted to each tree and location. And with a strong focus on flexibility and universal use(Also for disabled people).

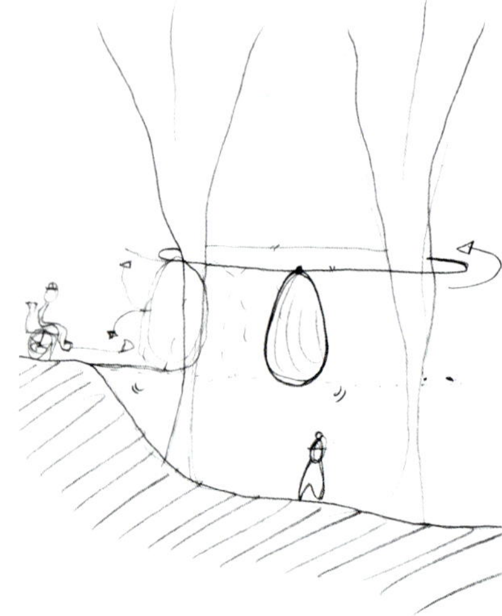

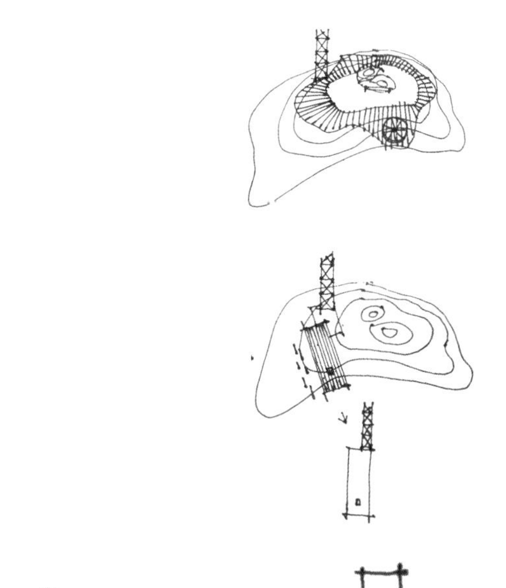

建造计划很快就确定分为三个部分：一个供使用者活动的平台（platform）、一个保护望远镜的覆盖物（covering）和固定望远镜支架的基础（base）。对无障碍设计、建造和使用需要的关注很大程度上影响了设计的最终结果。

由于时间紧张和场地条件复杂，我们放弃了用木材铺满整个基地，根据地形起伏形成一个宽阔曲折的地面平台的想法，而选择了 6m×12m 的平整宽大的矩形平面。这个尺寸的确定是基于轮椅使用的需要。覆盖物的设计也从建造一个较大的空间以供观测者使用简化为一个可移动的木盒，因为复杂的屋顶结构会给观测带来不便。使用的时候可以把它通过铺设在平台上的导轨推向一边。木盒本身也可以作为座椅使用。考虑到望远镜需要绝对的稳定性，基础被设计成一个直接固定在地面上，独立于平台的圆柱形钢筋混凝土块。

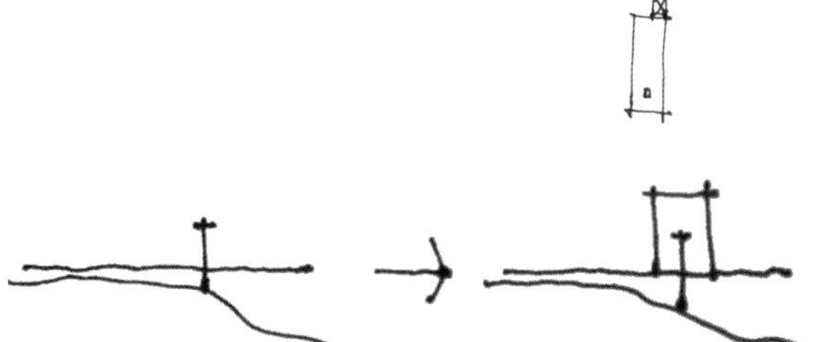

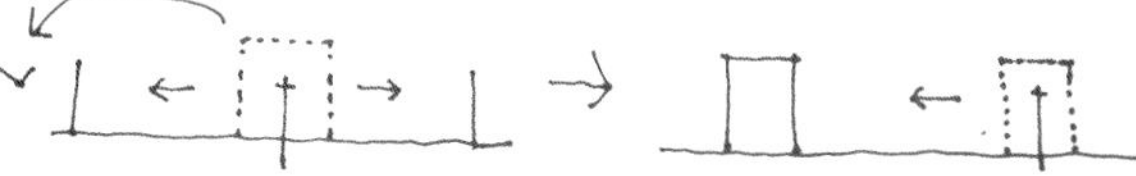

The construction program was quickly divided into three parts: a platform for various activities, a covering to protect the telescope and a base to fix the telescope on. The consideration for the disabled, construction and requirements for usage influenced the final outcome to a great extent.

The limited time and complex site situation obliged us to abandoned the primitive concept of platform that covered the whole site with woods to establish a artificial palyground with complex surface according to the site. Instead, we chose a huge 6m× 12m rectangular plan which was determined by the requirement for using wheelchairs. The concept of the covering was also simplified from a big shelter to a small movable wooden case because the roof structure would affect observation. It could be pushed to another side through the lane installed on the platform. It could be a chair as well. The base was a concrete column which was fixed directily to the groud and independent from the wooden structure for the stability of the telescope.

坚硬的山体岩石可直接作为地基，这为施工节约了大量的时间。由于地形起伏，确定立柱的高度须用水平仪根据平台高度逐一测量。在建造平台的主体结构的过程中，我们经常使用一些辅料以帮助矫正结构变形。

We didn't spend much time on handling the foundation because the stiff mountain rocks were naturally perfect for construction. As the complex situation of the surface of the site, we had to use the levels to figure out the exact length of every columns according to the height of the platform. We also frequently use some small pieces of wood to help rectifing the main structure while constructing.

剖面图 Section

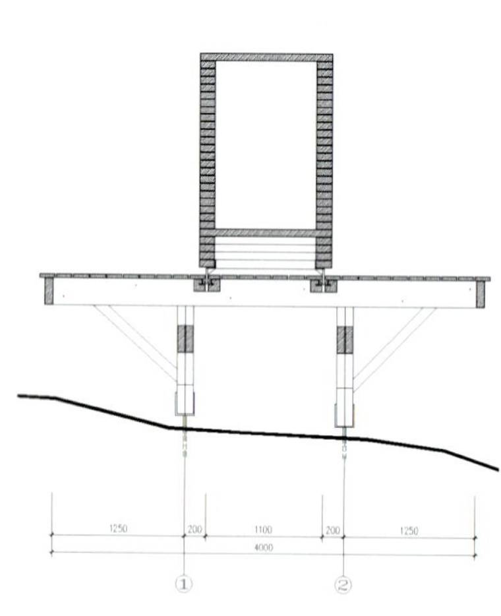

立面图 Elevation

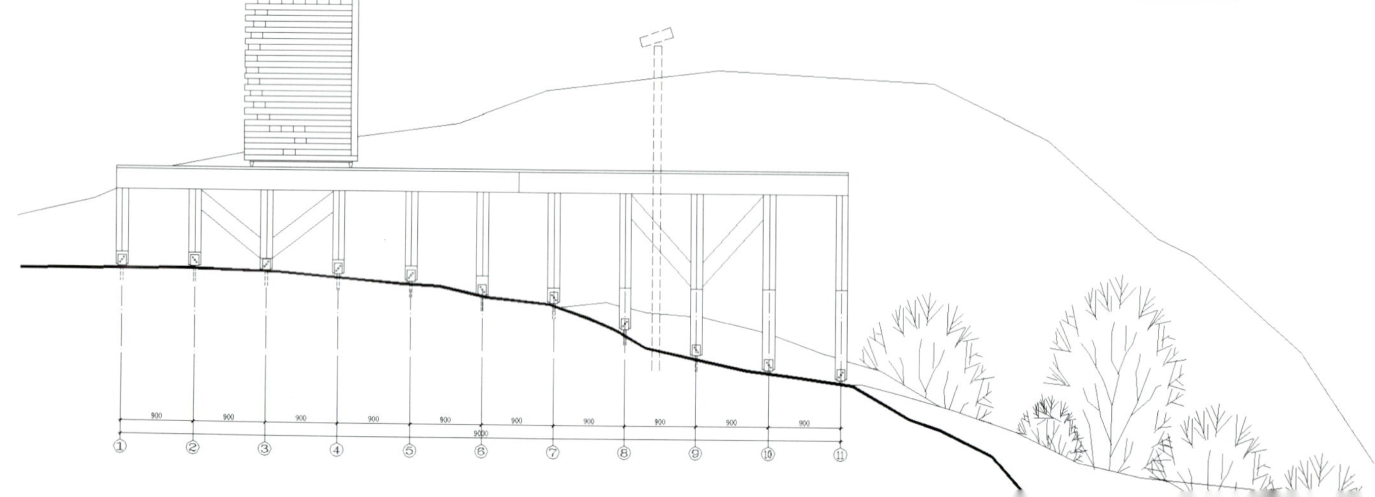

每天的工作时间很长，一般都要工作到天黑才下山。还好，山顶的风光很美，从这里可以眺望北海，郁郁葱葱间点缀着一些木屋。工作之余，可以欣赏一下美景，放松一下心情。

We worked for a very long time everyday. Fortunately, we had a perfect view at the top of the hill where we could overlook the North Sea. The site was surrounded by mountains, and little wooden huts interspersed in the boundless forests. Sightseeing was our best way to relax at the spare time.

由于基地在山顶，运料是一大难题。我们采用人工接力的方式从后山选择一条短而陡的小路将木料一根一根传至山顶。为此，全工作营的老师和同学甚至客户都参加了进来，这也成就了大家聚在一起共同劳作的有趣而壮观的场面。

Carring the timbers to the work place was a big problem because of the location of the site. We decided to carry the material one by one by hand through a short but steep path at the back of the hill. All the teachers and students came to help us, even the clients did. It was a very interesting and spectacular scene that all gathered and worked together.

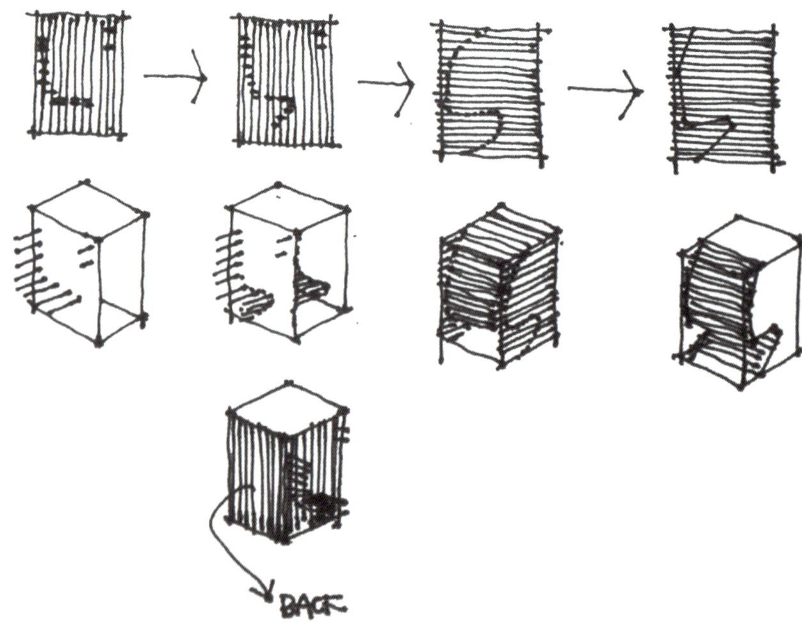
BACK

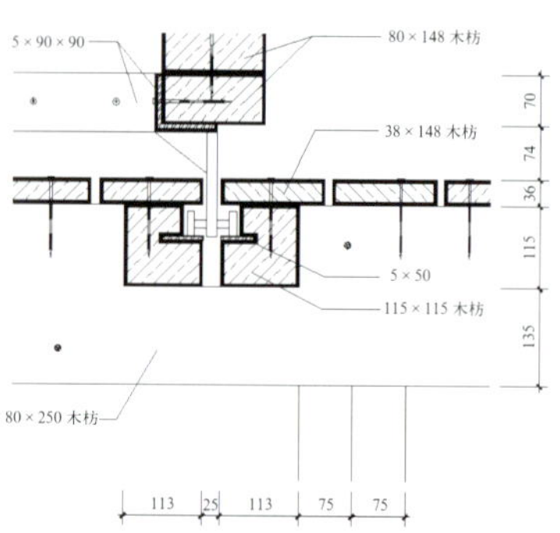
5 × 90 × 90
80 × 148 木枋
38 × 148 木枋
5 × 50
115 × 115 木枋
80 × 250 木枋
70
74
36
115
135
113
25
113
75
75

国际木构工作营（南京/特维德斯特兰德）建筑设计教师与建造匠师
Teachers and Masters of architecture design and construction in the Cross Cultural Workshop with Wooden Construction（Nanjing/Tvedestrand）

赵辰 南京大学建筑学院教授
Zhao Chen , Professor SA, NJU

冯金龙 南京大学建筑学院教授
Feng Jinlong , Professor SA, NJU

周凌 南京大学建筑学院副教授
Zhou Ling , Associate Professor SA, NJU

程云杉 东南大学建筑学院博士
Cheng Yunshan , Doctor SA, SEU

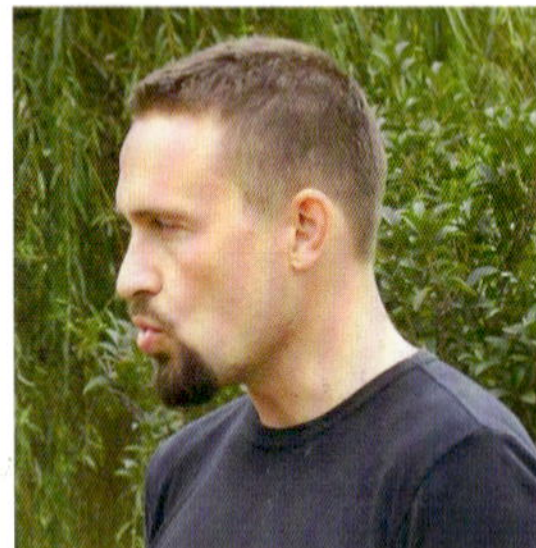
Sami Rintala
Professor II NTNU , Teacher AHO

Kolbjørn Nesje Nybo
Professor II NTNU , Teacher AHO

Odd
Professor II NTNU , Teacher AHO

国际木构工作营（南京/特维德斯特兰德）建造匠师
Masters of construction in the Cross Cultural Workshop with Wooden Construction Nanjing/Tvedestrand

何凤山 南京大学建筑学院木工师傅
He Fengshan , Carpenter SA, NJU

Torbjørn Seierstad
Client

Anders Lyche Oppegaard
Client

Jan Otto Gunleiksen
Client